THE TELEVISION INDUSTRY

UNDERSTANDING AND WORKING WITHIN THE TELEVISION INDUSTRY

Andy Hughes

EMERALD PUBLISHING

Emerald Publishing
61 Inniskilling Road
London E13 9LD

British Library Cataloguing in publication data. A Catalogue record for this book is available from the British Library.

ISBN 19031909 13 9

Cover design by Bookworks Ltd

Printed by Bookcraft, Wiltshire.

THE TELEVISION INDUSTRY
CONTENTS

Thanks to Teri and Roger for support, advice and endless patience – what shall we do now?

Dedicated to Mary, Tanya, Charlee and Frankie with love always.

1

INTRODUCTION

Ahhh television! The very word brings a warm glow to your heart.

This is where you belong, this is the mystical place where all your ambitions are going to be realised, and all your dreams are going to come true, and you are going to enjoy the career you have always dreamed of - at least that's the reason why you have chosen this book, and all those things could come true. But spare a moment at this early stage to check out a few of the basic facts about your proposed new home, the place that will embrace and reward your unique talents in equal measure.

How are you at rejection? Do you take the word 'No' badly? Do you crawl off into a hole and die inside for a day or so before you can face another possible rejection? What about being shouted at? Never acknowledged, never thanked, taken for granted, blamed for things that are not your fault, credit given elsewhere for your ideas and actions, generally made to feel like a useless dogsbody while you do your job well and consistently to the best if your ability and beyond, and maybe for little or no financial reward…

Is this putting you off? If it is, then television is not the place for you, so you'd best put this book back on the shelf and think about another career direction.

If you've already considered these possibilities, or are considering them now, and you don't feel in the least bothered by the prospect of this potential terminal career rejection hellhole, then you may be just the kind of person that the television industry is looking for.

The simple truth is, to get anywhere in television today, you need a thick skin, endless patience, talent and ability, a willingness to carry the can when it's not your fault, and get taken for granted when things go well.

It's not that television people are heartless ungrateful egomaniacs, well not all of them anyway, it's that television is a massively complex operation with large sums of money riding on the abilities of a large team of people to put the whole thing together and make it work.

To work in this industry, you have to be a people person and team player, and well able to shelve your own feelings and frustrations while others give vent to theirs. Your skills in your chosen area will be taken for granted – no-one is let loose on a television project unless they have either proved themselves as a consummate professional, or are in the business of providing that proof, and working for next-to-nothing, or even for free while they do so.

It's the indefinable attributes you need to be able to succeed in this industry. You will need to be a genuine (forgive the cliché) 'people person' because that is what's going to make you stand out and be remembered for the next time someone is choosing the staff for a project, and your career area comes into the picture. Yes, you are excellent at your job, but you also work well as a team, you are calm under pressure, you don't throw a tantrum if someone shouts at you, or forgets to say thank you, or asks you to re-do something for the tenth time, or is just generally obnoxious because they got out of bed the wrong side. Television is one of the most pressurised industries to work in because the people working there are always short of two major commodities – time and money.

You need to be able to work without an excess of either, but still produce a finished product that suggests you've enjoyed a surfeit of both.

If you still feel that working in television is the only place you are ever going to be truly happy, and you have read all the previous pitfalls and problems, and you still want to carry on and find your chosen job, then read on ... the rest of it is easy!

2

THE HISTORY OF TELEVISION

When you look at the volume, effect, and impact of television as we tentatively step into the new century, it's worth remembering the speed and volume of the evolution of television.

Ask most people who invented television, and you'll get the name of John Logie Baird, which is only partially true.

One George Carey from Boston conceived the idea of breaking down televisual images into individual components transmitting them via multiple circuits as far back as 1875, and an individual named W.E. Sawyer decided that images via a single wire would be possible by scanning picture elements.

With ideas such as this, television slowly but surely took shape and form.

The really significant developments in television technology began with the work of Vladimir Kosma Zworykin, a Russian-born American inventor, and credited with the creation of the iconoscope in 1923. Although this was designed for picture transmission, the iconoscope actually formed the basis for the first television cameras. In 1924 Zworykin filed a patent application for his kinescope, and by 1929 he was demonstrating the concept of television, although his screen was only one inch wide! In the particular way that television has of affecting those who come into contact with it, within his lifetime Zworykin was upset by what he saw as the 'trivialisation' of everyday life, and was far more interested in using television as a learning medium. Quite what Mr Zworykin would have made of the breadth of output of his invention today makes interesting speculation – a man who always wanted his vision used for human education and advancement seeing *Big Brother* or *Jerry Springer* or *The Simpsons* makes an unusual prospect!

In 1924 in England, the well-known John Logie Baird made experimental transmissions, providing a public demonstration of the new technology in 1925 from Selfridges department store in London's Oxford Street.

1925 saw Charles F Jenkins demonstrating 'radiovision' including 48 lines per inch, and synchronised sound over a distance of five miles. Baird began operation of a 30-line transmission a year later in London.

The television baton was passed to Philo Taylor Farnsworth who pondered the possibility of television in his early high school years. In 1927, at the age of only 30, Farnsworth transmitted his television image, broken down into 60 horizontal lines. With the kind of far-sighted vision that seems to have been a hallmark of the original television pioneers, Farnsworth's first image was a dollar sign.

Television technology refinement and innovation continued – by 1928, more than fifteen stations were licensed for television broadcasting, and the first receivers designed for home use were sold, with three-inch screens.

By the end of the 1920's Network Television was born, In London the BBC was sending out test broadcasts of 30 minutes per day, using the 30-line system created by Baird, and a mere two years later in 1931, CBS upped its output to 28 hours per week.

In 1939, the first commercial television sets were available, but as has been the trend for almost all developments in television technology, the first sets, the TRK-12 models, cost an astronomical $600, and it would be some years yet before televisions became a standard item in the homes of most people.

In 1948, Cable TV was brought into American homes for the first time, as Community Access Television. The entrepreneur in question was an electronic appliance storeowner named John Watson who provided television to the good people in the

Pennsylvania mountains. Provided they could afford to buy a receiver from his store in Mahanoy City, a hook-up to the new system was charged at $100 and a further $2 per month thereafter.

Moving forward to 1953, Peter Goldmark created the three-filter system that would allow television to be broadcast in colour. Goldmark holds a place in the hearts of television aficionados everywhere by assisting in the invention of the first electronic video recording system.

Throughout the 1950's and 1960's the increasing use of satellites for military purposes were soon to evolve into commercial possibilities. In 1962, Vice-President Lyndon Johnson took the first telephone call from space, followed quickly by television pictures from space in 1963.

A further landmark in the history of television transmission occurred with the launch of the Earlybird satellite in 1965, providing a satellite link up for various communication media, including television pictures, between Britain and the U.S.

The serious use of satellite television arrived in 1975, leading to the advent of subscription programming in 66% of US homes within ten years.

Trends were set by now – the first major telecast via satellite to a mass audience was the heavyweight boxing match, the much hyped 'Thrilla from Manila'. By 1979 Warner Cable had begun the Star movie channel and the Pinwheel children's channel, soon to be re-branded as The Movie Channel and Nickelodeon. The final piece of the major television audience was catered for when Nickelodeon took on the Popclips video show produced by Mike Nesmith, formerly of the Monkees, who had pioneered the notion of promotional videos for songs back in 1976 with the film to accompany his song 'Rio'. The vision shown By Mike Nesmith translated directly into the MTV format in 1981, and ensured that Nesmith enjoyed a third multi-million dollar fortune to add to the two he already enjoyed – as one quarter of The Monkees, and a successful musician in his own right, and as the owner of the patent for typewriter correction fluid which was passed to him from his mother, a former secretary, who came up with the original idea.

By the mid-1990's a further 75 new cable channels came on stream, joining the 78 existing channels, ensuring that the two major players in cable television – TCI with 10 million subscribers, and Time Warner with 7 million could increase their powerful grip on the medium, and that their revenue, which already exceeded that of the three major terrestrial networks, would continue to grow.

In the last decade of the old millennium, the advent of digital television saw an explosion in access, stations, programmes, channels, and most important as far as you the reader is concerned, jobs and job opportunities.

There has never been a better time to try and find a career in television, simply because there has never been such a large number of programmes being made, to satisfy a huge, and ever growing worldwide audience.

3

HOW TO GET YOURSELF A JOB IN TELEVISION

This is where the work really begins. If you've read this far, you already know that your future career path is something akin to sliding up a greasy pole while balancing on the point of a rugby ball and singing songs from musicals in a falsetto voice. If you've skipped straight to this chapter and not read of the potential pitfalls, it won't hurt you to be advised that getting into television is like getting a date with a film star, not impossible, but a long way from being easy and straight forward, so be prepared.

Disappointment and rejection are going to be good friends from now on, and you won't have to wait long to make their acquaintance, so try and avoid the earliest and easiest pitfalls by taking the advice that follows. There are a few tips that can help you in getting to where you want to be, avoiding some of the tears and tantrums that come with the territory.

THE LETTER

Whichever area of television you want to enter, the chances are your first contact with the fabulous individual who actually gives you some (hopefully gainful) employment, is going to be by letter.

The correct approach to your first contact letter is of vital importance, and it helps to put yourself on the receiving end of someone who gets hundreds of thousands of letters like yours – what would you like to read, and more importantly, what sends a potential hopeful's windswept-and-interesting missives straight into the shredder?

Don't be a gusher – someone who writes something like 'I'll do **anything** to get into television!' comes across as a fool. You have to show that you know the person to whom you are writing, and you know they are the right person because you've done your research in advance, and you are not simply scatter-gunning letters like confetti in a hurricane in the hope that something may come back to you in the form of a job offer.

You have decided your specialist area, you've tracked down the right person to contact, and you are out to impress them with your abilities and skills, so trying to advise a trained professional with limited time, and even more limited patience, that you are the person for them because you will 'do anything' means you are destined to be forgotten before your letter has hit the rejection pile.

What you need to avoid is the impression of enthusiasm bordering on psychosis, with nothing whatever to back it up. Sure, you need enthusiasm, your recipient has got where he or she is because they are skilled and talented, and they eat sleep and breathe their particular area of expertise. You need to prove that you are as keen as they are, but more importantly, you can back up your willingness to work with some experience and qualifications that will make you an attractive prospect as an employee. Don't be keen to bang on about 'learning at the feet of the master' or some such similar nonsense – yes, television is powered by egos, but false flattery to a complete stranger is likely to provoke evasive action rather than a smile of delight.

Make your letter business-like, and as short as possible, while including all relevant information. No point telling a prospective employer that you've always wanted to work in television, this can be understood on the basis that you are applying for a job in the first place.

Your letter should be word-processed, double-spaced, addressed correctly, including appropriate contact details, with a separate CV showing all relevant qualifications and attributes. If you haven't got a professionally prepared CV, then obtain one, it will cost a few pounds, but it is money well spent – presentation is vital at the first stage, so spend the time and a little money to make yourself look good. Remember, if you can't type a proper letter, and present a properly prepared CV, you are already doomed to failure, so you should be acquiring basic computer skills if you don't already have them – they are taken for granted in just about any area of modern television. Check you've spelled your recipient's name correctly, and never begin with 'Dear Sir or Madam' which infers that you are writing to dozens of people simultaneously, or you can't be bothered to find out the name of the person you need to contact, both of which can assist your letter towards a more useful destination in a pulping mill somewhere!

Ensure that you conclude by advising that you are available for personal interaction at any time, and any place, and make sure that you are available.

THE C.V.

Almost as important as your initial letter is your CV. It's surprising how many people cobble together a mass of useless information that goes on for pages, and remains unread past the first few lines. Spend a few pounds getting a professional CV prepared – your local paper small ads, will probably have advertisements from people who compile CV's, so go and see one of them and get the job done properly. It's money well spent because presenting career and qualification information succinctly and coherently is a specialist skill, so don't try to do it yourself. Any prospective employer will see at a glance that your CV has been professionally prepared, and it's a point in your favour that you've taken the time and trouble to present yourself properly.

Your CV must include all relevant experience, and you need to have as much of that as you can get. Of course, your experience is limited at this stage, but an absence of professional experience should be compensated by your proven willingness to learn what you can where you can – you should be signed for any appropriate training courses to which you have access, and any unpaid experience in your chosen field will show a willingness to work hard to get somewhere, rather than the often received impression that the world of television is a poorer place without your good self within it.

A properly presented CV will always enhance your chances of proceeding to the next stage in the recruitment process.

FIRST CONTACT

Your letter and CV have made it through the 'paper sift' process, and someone actually wants to find out more about you.

Because your contact details are clear and accurate, your letter reader will have had no trouble in writing, phoning, or e-mailing to you, and of course you pick up their response instantly and reply as soon as physically possible. Don't confuse a prompt response with the pointless 'I'll do anything' approach outlined earlier – this is you providing the courtesy of a prompt reaction to someone who has taken the time and trouble to investigate you further as a prospective employee. Failure to respond instantly will mark you down as lazy or discourteous, or inefficient, all of which will ensure that your employment prospects have taken an irreversible turn for the worse, so make that return contact quickly.

If your letter reader is merely writing back to give you a 'thanks but no thanks' brush off, always reply with a brief and polite 'thanks for your time' note –it costs little, and leaves the contact with a good feeling about you – which just may stick in their mind in the future. Of course, it may not, but good manners are still free, and it stops you getting bitter and twisted!

If, on the other hand, you receive an offer of further contact, be ready and willing to swing into action.

You may receive a phone call, in which case be ready with a few answers to obvious questions – why do you feel you are suited to this job? What skills and abilities would you bring, and so on and so forth. Remember, to speak clearly, don't waffle, and don't be afraid to finish your answer and wait for the next question – this isn't a dinner party, so it's not your responsibility to keep the conversation going.

If you receive a letter or e-mail asking you to come and meet your prospective employer, then you agree to their chosen time and place, and follow the previous rules.

You have to sell yourself, but that doesn't mean bragging and being clever. Remember, you will be called upon to prove your skills and techniques which are the reason why you are being employed – no-one likes a timewaster, so ensure you can do what you say you can, or you will look like an idiot, and an unemployed idiot at that!

EARLY STAGES

Never forget you are the new kid on the set, and be aware that a certain level of ego is a prerequisite in the world of television, and some people's egos are the size of Jupiter. If they happen to belong to the people to whom you are responsible, you will need to learn the art of tact and diplomacy. Remember, there are no passengers in television, so however unpleasant someone's personal manner may be, they possess experience and ability which you can learn from. Nature gave you two ears, and two eyes but just one mouth, so use the first most, the second almost as much, and the last as little as possible. Behave like a professional from day one, and eventually you will be treated, and more importantly, regarded and paid as a professional. Obvious advice? Maybe, but watch as others around you ignore it, and pay the price accordingly.

4

BEHIND THE CAMERA

Plenty of people would love to 'work in television', but they have next to no idea of what jobs are there are, and what skills and qualifications you need to get the job that will suit you best.

So, you sit there day after day, night after night, soaking up all the TV you can get your eyes on, and when the credits roll, you look at the job descriptions, and the lucky individuals whose names are next to them, and maybe you wonder just what exactly a grip' does, and what a 'rostrum camera' camera is, as well as the more obviously descriptive job titles such as 'director' or 'costume designer'.

This is the part of the book where some of the occupations that make up a television programme are explained, and you can decide if you are keen, and indeed able, to provide the attributes needed to be good enough to get your name on the credits of a programme of the future.

Work experience is the best place to begin any television career. Yes it's probably unpaid, and you get precious little thanks for flogging yourself silly, but it's the way to shine above the many who dream of a career in television, and to become one of the few who actually finish up with one.

So, you are the new kid on the block, how do you make an impression? Arrive first, leave last. Be willing to do anything asked of you. Keep your eyes and ears open and take in as much information as you can.

You will probably be working with the Runners and Researchers, which are effectively the next steps on the ladder you are attempting to get a foot on, so be very nice to them, and offer to help as much as you can. Try to ensure that you offer your services to a company that makes the kind of television you want to work in eventually. It's pointless spending your days on nature programme locations when you are more interested in the internal workings of a combustion engine – so aim roughly for the area that will suit you once you've found more precisely what you want to do.

When your stint ends, send a thank-you note to the Producer you worked for, and also to anyone who was especially helpful, and include your contact details. You never know, when another post comes up, if you've made a good impression, they will remember you. The posh name for this kind of activity is 'networking', but it's really just common courtesy combined with common sense. If you want people to remember you, give them your time and energy when you are there, and your contact information when you are gone. Simple, but effective.

RUNNER

The best starting point if you are young and enthusiastic, and you don't want to be paid in cash, but more in experience, then being a Runner is a great way to see what goes on, with little of the actual responsibility of making it happen.

Runners, as the name implies, are there to fetch and carry for other staff who are too busy or too important, usually both, to do their own 'running', and yes, you may literally be the tea-boy (or 'tea-person, but don't make an issue out of it!)

Being a Runner gets you on the inside of a television studio, or a location shoot, and lets you see the nuts and bolts of television as it is made. You can see what other jobs exist, talk to the people who make television happen – but be careful to pick your moment, whispering to the Floor manager about his or her job as they are counting down a Presenter on a 'live' show will not enhance your future, or indeed your present career prospects. Break and meal times, when people are relaxed, is chance to listen and learn, and ask questions of people who appear friendly and have a few spare minutes. Most important of all, be on time, do what is asked of you, remain unflappable, even when you are being shouted at for something that is not your fault, and above all else, **be reliable**.

This will ensure that you are noticed and remembered, and when the possibility of a job in your prospective area comes up, hopefully someone will think favourably of you. Contacts are everything in television, it is very much what you know, but there is a fair degree of who you know, especially when you are making your start. Never miss an opportunity to get your face and your name noticed, but don't be pushy about it – telling people you are the vital cog their production is missing will simply irritate them, and media professionals don't take kindly to being told what or who they need to make their programmes a cut above – it's a short cut to a new career for you – probably outside television, which rather defeats the object.

RESEARCHER

Another essential area of television, but one in which staff contract posts are rapidly disappearing, so most Researchers work on a 'programme-to-programme' basis, or maybe short term contracts of maybe six months or a year.

You should do your research (no pun intended) in advance before you approach a Producer, or a Company for work as a Researcher. You will need to be able to work methodically, at short notice, and often under pressure. Researchers tend to be given tasks and left to get on with them, so a large degree of initiative and self-reliance are essential here. Producers and Directors want results, not excuses, you produce the maximum on the former, and minimum on latter, and accept that lack of gratitude as part of your job description.

Being a Researcher is about ideas – so keep an 'ideas file' – either notes, or properly thought-out programme projects, and you can use and add to it as you go along. Try to keep your mind open for programme concepts based on what you see around you. Programme ideas come from people who ask why and where and how. If you are sitting in at a production meeting, see if you can add a few ideas to the melting pot – Researchers are not people who sit around and say nothing, those that do will be seen as stupid or secretive, neither of which is an asset to your future career.

Flexibility, such a key word in most television careers, is a virtual template for a Researcher – anything from taking a self-operating camera and sound-system out to a location to provide a busy Director or Producer with some on-site info. and preliminary material, to booking pop and film stars for a chat or magazine programme, the tasks are many and varied. You will need basic keyboard skills, and knowledge of the Internet is increasingly important.

Most people who become good Researchers stay in that job because it suits their particular skills and they become good at what they do, but as with all occupations, it is possible to move onto other areas in television if you show the aptitudes and abilities required.

CAMERA OPERATOR

One of the most obvious television careers – television would not exist without the cameras that create televisual images, and therefore one of the most crucial. A camera operator needs a good depth of technical knowledge – when you turn up for your interview, even at work experience level, and they ask you what is meant by 'depth of field', they expect a suitable answer not a blank look and some panic-induced waffle. Knowledge of lighting and sound, and an inbuilt instinct for interpretation of the director's ideas, together with the tact to correct his or her mistakes without making it obvious, a large reserve of physical stamina and concentration, all are essential to the camera operator. If you think you might vaguely like to be one of those guys you see swinging around TV sets with headphone on, forget it now. Unless you have an inbuilt passion for images, and staging, and getting the best visual presentation in any situation, you don't have the basic skills needed, so look somewhere else. If camera work is for you, look at professional qualifications and courses available in the appendix to this book and take it from there.

FLOOR MANAGER

One of the most skilful and stressful jobs in television – definitely not for the faint-hearted! In the theatre, the Director works from the theatre floor, and is visible to the actors and crew during rehearsal, only retreating out of sight for the performances. In television, the Director works from the gallery – a control room linked to the studio floor by intercoms and sight monitors. Because the Director cannot see the entire floor at any time, he relies on the Floor Manager (sometimes the Second Director on larger productions) to ensure that his instructions are complied with, and that everyone is where they are supposed to be at any one time.

The job is especially stressful on 'live' television, where there are no opportunities to re-shoot scenes, so the Floor manager has to be on top of his job all the time. It is a job that requires excellent people and communication skills, a sound knowledge of the technical aspects of television production, and the ability to juggle several lots of instructions at once, while keeping cool and in control. A number of television Floor Managers come into television via the theatre, where they gain valuable experience in running a co-ordinated team of staff.

EDITOR

Editors work in what is called 'post-production' – taking all the disparate ingredients of a television programme and knitting them together in a cohesive whole. Because the work involves working in an editing suite, most Editors come into television via technical editing courses, or specialist University degrees. The Editor's job is to realise the Director's dream, and make the best use of the material available while ensuring close liaison with the Director and Producer means that the original concept is maintained. It's a job which involves artistic instinct, married to technical ability, and again, a good dose of people skills to deal with Directors who are convinced that they have given material of pure genius for you to work with, and will be less than impressed if pure genius does not shine through the finished result. Reaching a suitable compromise between vision and reality is a large part of the Editor's skill, and it's a constant learning curve, with good Editors being virtually assured a job for as long as they want it.

PRODUCTION MANAGER

When you look at the long list of people involved in making even the most basic television programme, do you ever wonder who's job it is to ensure that all these people are available at the right time, and in the right place, that they all get paid, fed, given beds for the night, have transport, the props and costumes al arrive, and there are places for people to change and make up, that all the technical equipment is in the right place, and everything is in place when the filming starts? Then maybe a Production manager is the job for you. The inevitable people skills will assist, and you have to have a very organised mind with infinite attention to detail. Making television is an extremely costly exercise, and no one will be too impressed if the cast and crew break for lunch and there is no lunch because you're on location miles from anywhere and you've forgotten to bring in mobile catering facilities. A Production manager's job is very much 'behind-the-scenes', but is certainly no less vital for that.

SOUND ENGINEER

Another specialist 'behind-the-scenes' career that is suited to a particular set of skills. Sound people are utterly passionate about sound – they know how to ensure that an actor can be heard clearly even though he is standing in front of a boiler factory in the middle of a thunderstorm, with a dog barking in the distance, and a train passing by. The mixture of these sound levels to ensure that no one interferes with any other is the Sound Engineer's job, both while shooting, and afterwards during the editing process.

By definition, these people are passionate about sound, and equally keen on the workings of the technology that records and reproduces it, so a science background and a willingness to work with vagaries of nature, and directorial vision are essential. Perfect hearing is taken as read.

PRODUCTION SECRETARY / P.A.

Although the more snobbish individuals in television (and there is no shortage of them!) see secretarial people as menials, they are every bit as essential to a production as any of the technical and front-of-camera people.

A Production Secretary will probably have begun his or her television career as a secretary in the production offices, and already possess excellent communication, keyboard, and interpersonal skills, but making the move to a Production Secretary calls for an additional set of talents, as well as ensuring that the precocious ones are tested to their limits.

These are the people who get things done – they move heaven and earth to get particular tasks done, often taken for granted, but still vital, so if you are looking for job satisfaction, rather than constant assurance of how wonderful you are, then this could be a career area for you to explore.

Production Secretaries and P.A.'s need to be able to think on their feet, and have answers to problems before they arise. One of the tasks of a Production Secretary is to time everything that is filmed, and maintain an up-to-date log with all the information instantly to hand – think of it as being able to add up like Carol Vorderman, and fire off answers like Bob Monkhouse, and you get the general idea!

5

IN FRONT OF THE CAMERA

Lots of people who want to work in television want to be the 'personality' – someone who is a household name, known everywhere, asked for autographs, and invited to all the swish showbiz parties. Sound like you? Maybe it is, but be aware, like everything else in television, there is huge competition to get a very few posts, and it's not as easy as it looks!

TELEVISION JOURNALISM

Quite a number of today's quiz and chat show hosts began their television careers as journalists. Michael Parkinson was, and still is, an excellent sports writer who has diversified into television and radio both in the UK and Australia. Parkinson has managed to utilise his journalistic skills – curiosity, interest in people, and ability to carry out detailed research, and utilise his skills as a television interviewer. In fact, Parkinson's skills make his task look deceptively easy – being relaxed and chatty with complete strangers, and famous strangers at that, is nowhere near as easy as Parkinson makes it appear – witness the myriad others who have attempted to follow in his footsteps, only to find their time at the top as difficult as it was brief.

Plenty of journalists are happy to remain journalists, reading news, writing news, working as foreign correspondents, or arts pundits, and maintaining close links with their original areas of expertise. Others, by default or design, find a career in entertainment opens up for them, but it is a rare combination of skill, talent, and good old fashioned popularity that puts you at the top, and keeps you there.

Beware that having no private life, an intrusive press checking into your life, as well as that of your family and friends, and the constant heat of rising stars breathing down your neck and looking to step into your shoes before you even stepped out of them, calls for a particularly tenacious personality – make sure you want what you dream – you may get it!

ACTING

Acting is a profession you should only consider if you are going to be utterly miserable doing anything else, and for television acting, you can triple that motto.

The obvious route into television acting is via theatre – without experience and an Equity card, you will have next to no chance of being employed in television. A number of actors find they prefer the challenge of live theatre, others find television work is their natural home, and become the 'bankable' stars for their employers – people like David Jason, Amanda Burton, Martin Kemp, Pauline Quirke, actors who's name above a title will guarantee a large audience.

Television acting is almost the opposite of theatre acting – on stage you need to project your voice and exaggerate your movements, make your presence fill a large or small auditorium. In television, the auditorium is the camera pointing at you, and your skill becomes the smallest nuances you can project with your eyes, or mouth, a glance, a small body movement, these become the tricks of your trade. Television actors bring a natural grace to their work, if you can't see it, they are doing their job well, but you can always see when the role is forced or stilted, the camera is utterly merciless, and you need those skills to get, and keep, any acting job in the television industry. Lucy Davis, who plays 'Dawn' in the massive hit comedy *The Office*, provides a perfect example of television acting. Her dialogue is less vibrant than some of the other characters, but her conveyance of her emotional turmoil over her attraction to 'Tim' is conveyed entirely by glances, and eye movement. The essence of good television acting is the ability to make 'small' into 'large' and in this role, Lucy David demonstrates that skill to perfection.

Some people who have made their careers as comedians and entertainers in clubs and theatres go on to work in television – Des O'Connor and Shane Ritchie both began their careers as holiday camp entertainers, the routes into television are varied, but there is a large degree of luck and 'right place right time' involved, to say nothing of a large amount of persistence.

Increasingly, television actors and actresses are finding a popular role is a stepping-stone to every higher profile roles, and in some cases, a chance to step into the worlds of television outside that of straight drama or comedy. This can be a two-edged sword – witness the effect of a show like *Celebrity Big Brother* the 'reality' TV show (a contradiction in terms if ever there was one!) on the careers of actress Clare Sweeney, and quiz show host and actor Les Dennis. The former found herself catapulted into a West End stage musical, and the potential to follow her fellow Liverpudlian Cilla Black to the very top of the tree. The latter found his eccentric behaviour the butt of tabloid jokes, his already shaky career virtually wiped out, and his on-screen confessions hastening the end of his marriage.

Extreme examples perhaps, but being a 'personality' means that people feel they know you intimately, even to the point where they will pull up a chair at your restaurant table and start advising you that the suit or dress you wore on television last week really wasn't 'you'! They're quite right of course – it was chosen for you by the wardrobe department of the studio at which you were working, but your public will feel it is quite within its rights to have a moan at you, even if you are out with your family for a quiet Sunday lunch. Again, the inevitable conclusion arises – if this sounds scary to you, then don't aim for this career, you won't enjoy it, and you won't last

– the public can spot a fraud a million miles away, and your fall will be merciless, and very public. Avoid!

6

CAREER OVERVIEW

A major reason cited by people who want to work in television is the 'glamour' of the job. This is usually a perception created by what we see when we switch on our television sets – bright witty attractive people being bright witty and attractive in exciting places, with famous people, getting paid a fortune – who wouldn't want a job like that!

Of course, television is an iceberg, and this is the one-tenth that the viewer is aware of, and the remaining 'behind-the-scenes' tasks are what take up the bulk of this book, and we'll get to them shortly, but let's take a look at the 'glamour' frontage of television, and see it for what it really is.

Turn on to watch Saturday morning television for example, and you will see a slick hip young presenter talking to teenagers and pre-teens on their level, and making it look as easy as falling off a log. And getting paid loads for it as well!

Yes they do, because what you are seeing is possibly the most difficult and skilful job in modern television – performing a function that is as fraught, technical, and downright terrifying as it appears easy, simple and fun.

Live presenters are the lynch-pin of the whole thing, and the potential for them getting it wrong and looking stupid are large enough to give the average viewer a migraine just thinking about it, never mind dealing with it when it happens.

You see the hip young presenter (the HYP) chatting away as though he is talking to just you, and you are best mates. Easy ... except that through the tiny earpiece all presenters wear, there can be a constant chatter of instructions. Instructions to the presenter – no problem, except that to save the job of tuning everyone's earpieces in and out to get specific instructions to specific people, everyone hears everything for everybody, because it's easier that way. So, a babble of voices giving instructions to maybe a dozen people, cameramen, lighting technicians, taper operators, the floor manager, are all coming into HYP's ear continuously, and he has to listen to anything that is aimed at him, and tune out anything that is for someone else, and all the time he has to keep talking sense to you, the viewer, and making sure he times what he says to fit in with studio presentations, live bands, adverts, and so on, and that doesn't allow for incidentals like 'The videotape just broke, speak for another twenty seconds can you …' just as his autocue (script rolling on a glass screen in front of the camera which the HYP reads out loud) has suddenly ceased to function and gone blank..

But they carry on seamlessly chatting and smiling, and the only time you will know something is wrong is when that rabbit-in-car-headlights look comes over the HYP's face, and he bursts into tears and runs off the set. Of course that doesn't happen, because the professional skills involved in this job are akin to piloting a moon rocket, sliding down a greasy pole, and carrying a tray of cut glass champagne glasses on a ship deck in a force 10 gale – all at once! Glamorous? Don't you believe it!

As with the introduction, if you were planning a career in front of the camera, and this has frightened you off, then it wasn't for you in the first place. You need nerves of steel, and the ability to keep calm in a crisis far more than you need to squeeze into a size eight dress, or wear the latest designer gear and look good doing it. Live presenting is a spectacularly difficult job, that is made to look childishly easy by the people who do it well, but the skills are as vital as they are unobtrusive, and you need all of them to make your way in this most visual area of the most visual medium.

If, on the other hand, you prefer to opt for a life behind the scenes, and that is where the work is done that makes the whole edifice come to life in the first place, there are a vast array of careers open to you, some of which may be obvious – Actor, Cameramen, and some far less obvious - Production Assistant, Floor Manager, and others, all of which have been explained in previous chapters.

7

MAKING A DOCUMENTARY

One of the ways to get yourself started in your proposed brand of television expertise is to join a team making documentaries.

In the past, it was always the norm to record documentaries onto film, because the finished product would usually be broadcast to a cinema audience. These days, documentaries are the domain of television, and the medium used to record them has transferred from film to video, partly due to the increasing sophistication of video recording, editing and transmission equipment, and additionally to encompass the varying range of locations used, and the costs involved in the programme making process.

Documentary subjects vary, but can be assigned to the following major subject areas - Current Affairs, Travel, Science, Safety, Public Relations, Adventure, Recruitment, Sales, Instruction, Wildlife, Art, Sport, and Social Problems.

It makes obvious sense to investigate the major channels, or the independent companies producing documentaries, and ensure that you aim yourself towards the companies that make programmes in an area in which you have either knowledge or interest. If you have spent your school days relishing the time spent in the chemistry lab, it won't really assist you, or your prospective employers, if you find yourself travelling to the far corners of the world to make travel documentaries.

The documentary is an ideal starting point for those interested in writing for television. **Script Writing** is a highly skilled job, and successful scriptwriters are in demand, particularly those who can adapt their style to fit the remit of individual programme makers.

Documentary scripts need to fulfil the remit of the programme, adhering closely to the subject matter, and simultaneously making the programme engaging and stimulating to the viewer. This rule applies to every type of documentary, from the most specialised academic audience, to a potential worldwide television audience encompassing a variety of backgrounds, cultures, and languages.

The 'write what you know' rule applies here, at least in the early stages of your television career. Of course it is possible to research and write about any subject available, but why make life difficult for yourself? When starting out in television, you need as many advantages as you can get, and knowing your subject, or at the very least enjoying an affinity with it, will assist you when the going gets tough. Discussing the script with the Director and Producer will expose any gaping holes in your subject knowledge, and leave you looking and sounding unprofessional.

Finding a niche in this particularly specialised area of television may mean looking outside the major television companies, and investigating the smaller freelance companies that increasingly supply material as the numbers of stations and networks continues to expand throughout the world.

Writing about particularly specialist subjects may well lead into researching and even directing, if only because of a lack of experience within the company for whom you are working. These opportunities should be exploited in full whenever they occur – no one ever failed to work in television because they knew too much about a subject in which no one else had even an inkling. This is a time to gently but firmly push yourself forward, be tactful, but ensure that everyone knows that you can fill that particular gap in the overall scheme. Remember, time is always money in television, if you can save both, you are on the way to becoming noticed, known, and eventually valuable.

A proven entry point into the documentary area is as a **Researcher** because it's a skill in demand for specialist programme areas, and within smaller companies, where more people tend to be multi-skilled, this can lead on to writing, and possibly even directing and producing.

A documentary will usually require a **Production Manager** and the route into this career can be followed through secretarial and Personal Assistant positions. Beware, this is a tough and demanding job, and it is not for the faint hearted! Production Managers have a powerful and responsible position – the job involves overseeing the production, including the overseeing of the production budget, the checking of facilities and equipment, in fact generally ensuring that all the 'nuts and bolts' areas of television programme making are taken care of, so that the rest of the team can get on without worrying too much about arriving at a location to find there is no accommodation, no catering, no camera crew, you get the idea.

Nothing much gets done in a television programme making without a **Camera Operator** and this is one area where simply being keen and willing is just not enough. It's impossible to bluff people into thinking that you can light and film a scene for a documentary, so you need to have attended the appropriate training courses and to have a flair for 'seeing' a shot through the director's eyes.

In charge of getting the ideas and visions down and onto film is the responsibility of the **Director** which is probably the most stressful job of all in television. Working with all the restraints of budgets, time, egos, and unforeseen difficulties, which never arrive singly, always in groups, the Director has to ensure that the programme is filmed and edited to the correct length, ensuring that all the relevant points are included, that no additional and useless material has been created at unrecoverable expense, and that everyone is happy with the end result.

The world 'pressure' could have been invented for the Director's position, but those who are good at what they do are ensured a long and successful career, so it is worth persevering if this is where you feel your talents lie. Visuals are all very well, but television documentaries are nothing without sound, and that's the job of the **Sound Engineer**. Those who love sound, and will not be happy doing anything else normally populate this specialist area of programme making. Because of the skills required, it is not an area the average television wannabe can drift into. Sound engineers develop their passion for sound at an early age, and are rarely interested in anything else. A love of technical wizardry, combined with the desire to ensure that the sound is the best achievable under any set of circumstances, marks out the potential sound engineer, and once again, the best are ensured a lasting career, because good sound is the lifeblood of any television programme, and those who can provide it are valuable indeed.

Taking the whole mass of creativity and nurturing it into the half hour / hour / series of programmes that can be presented to a television company to enjoy, and hopefully buy and broadcast, is the task of the **Editor.** To perform this vital function well, you need skills of diplomacy, the desire to tell a story, a feel for pace and an ability maintain interest – all of which you will need, in addition to the technical skills usually acquired during a course of Media Studies. Why a course of Media Studies? Because you will need experience of working in a professional editing suite, and access to those for learning purposes is usually provided on a recognised training course. Once you have learned the tricks of your trade, you will need your other skills to guide the Director through what he or she *thinks* they have created, into what will actually work in television terms, and still fit in with the directorial vision – it's not easy, but it's worth striving for. The end result will have the Editor's mark on it throughout, highly satisfying.

This is a basic run-down of the Documentary Crew and how it works together. Some larger crews will have additional staff, the smaller crews may have to double up in some areas, but all these functions will have to be provided for, and it represents an excellent starting point in television. For some, the documentary becomes their chosen profession, and they spend their whole working lives in that specialist area, but for others, it can be a solid grounding in television skills that can lead onto jobs in the other various areas of the television profession.

8

ANIMATION

Animation in the year 2002 is very sophisticated indeed and has grown into a very profitable sector. The name Walt Disney automatically springs to mind. There are many courses in existence specifically teaching the art and skill of animation, which will be outlined later. Animation – whether produced by the traditional well-known method of drawings, with models, or whether produced, as often is the case nowadays, with very sophisticated computer systems – is a very costly and labour intensive business. For example 30 minutes of animation, TV animation, may cost £350,000. Animation series may cost between £6000-£10000 per minute. And programmes can take many months to make.

However, it is the case that animation is becoming more popular as a form of entertainment and it can be marketed globally with greater ease as it 'travels well'. Animation has a long shelf life, dating far less quickly than other medium. There are regular showings of 1930's animated cartoons on TV and cinema with no sign of diminishing popularity.

If an animation production company gets it right, the long-term results can be very significant indeed. However, there are enormous risks. So therefore any undertaking has to be handled professionally and with great skill. Similarly with film, as we have seen, technological developments and the increasing need for product from the ever increasing army of TV stations, have encouraged new animators to challenge some of the older, more traditional, ways. A new wave of adult animation has reduced costs through clever scripting and other devices. Commissioning editors and schedulers in broadcasting are starting to become more flexible in their approach to funding and broadcasting. The reason is that animation can be very popular and ratings can justify decisions.

Animation generally

The UK animation industry leads the way in creative and artistic talent in this sector. A diversity of products include:

- Commercials
- Music promotion videos
- Children's programmes
- Feature films
- Adult comedy
- Adult series
- Adult drama
- Games
- Multimedia

Animation can be divided into two quite distinctive areas, although there is a lot of crossover. There is the traditional hand drawn 'cel' animation, the type most popularly associated with Walt Disney (the original). Every movement and facial expression is drawn and then transferred onto a clear cellulose for filming.

Between 12 and 25 drawings will be consumed every second, and up to 25000 drawings required for a half hour episode. Another development of the traditional approach, as typified by Wallace and Gromit is the clay animation and model animal characters such as those from Cosgrove Hall. There is still painstaking attention to detail but by manipulating models as opposed to drawings.

Computer animation is becoming ever more dominant with increasingly more sophisticated software coming into the market place. This sector can be very expensive and time consuming – with recent feature films using 400 state of the art computers. The challenges with computerised software are significant, as computers still cannot produce the same results as human beings. However, techniques are developing very fast and it is probable that computers will replace more traditional forms as time marches on.

Many new productions are now taking advantage of opportunities to mix cel or model animation with computer generated additions.

Jobs in the industry

A wide variety of people are employed within the modern animation industry. We have already seen that the film industry employs a very wide variety of personnel essential to bringing a film into being. The animation industry is similar. Far from the traditional idea of one person and their drawing board, again popularised by Walt Disney, there is an increasing emphasis on teamwork and multi-tasking. As production lead-times increase and budgets have to be managed, it is cost effective if people can move between projects at any one time. There are still plenty of people who work alone, with traditional animation requiring long hours of concentration. However, this is offset by the need to achieve goals within tight deadlines.

The following are some of the job titles to be found in the animation sector: After having read the previous chapters on film, many of the jobs will fall into place without further description:

- Animation Director
- Art Director
- Audio Producer
- Cel painter
- Character designer
- Clean up artist
- Designer
- Editor

- Graphic Designer
- Model Maker
- Music Composer
- Producer
- Rostrum Camera person
- Special Effects Animator
- SFX Rostrum Cameraperson
- Tracer
- Animator 2D-3D
- Assistant Animator/In Between
- Background Artists
- CD-Rom Producer
- Checker
- Colourists
- Dubbing Co-coordinator
- Games Developer
- Key Animator
- Model Set maker
- Post-Production Co-coordinator
- Production manager
- Set Maker
- Scriptwriter
- Storyboard Co-coordinator
- Web Designer

There are also a variety of lesser know, but equally important, jobs:

- Administrator
- Human Resources manager
- International Sales advisor

- Rights Specialist lawyer
- Specialist Accountant
- Fundraiser
- Investment Advisor
- Marketing Director
- Runner
- Studio Manager

It can be seen from the above, that the whole process of producing an animated film is every bit as complex as producing a feature film. In many cases, the process is far more time consuming. However, the business of animation is booming with the industry very positive about the future. International ventures are becoming more commonplace. There is potential for long-term earnings and there are added advantages, for example:

International sales – animation travels well and can be dubbed easily into many languages
Licensing and marketing – characters can become sought after brands which can be used to sell a vast array of merchandise, producing money for the animation company through licensing agreements.
Videos- the children's market in particular lends itself to video production. Sales, rental or archive material can produce huge revenues.
Compilation – the material can be reworked and used in different ways to produce additional income.

Financing animation

We saw with feature films that financing is never easy and quite often investment will come from a wide range of sources. Animated films, unless very short, are expensive and money often comes from the big studios. Disney has dominated the market for many years but other studios are getting in on the act. DreamWorks, Universal and Fox are increasingly active. Aardman (Wallace and Gromit) have made their first full-length feature film with help from overseas. Budgets of $10 million are not uncommon so it is obvious that a major player is needed for support in animated ventures of any significance.

Making an animated feature

As with feature films, there are a number of clear steps involved in the production of an animated film. They are as follows:

- Original story idea, based around a central character, sold to a production company or an international distribution company which supplies production finance
- The script is developed
- Animators developed the characters and backgrounds to fit the feel of the production
- Actors chosen for the voice of each character
- Animators produce work in line with storyboards/time-plans
- Sound track recorded in 'sound dead' booth
- Detailed design check on drawings for international target markets
- Continuing research to ensure that background looked OK
- Animatic – where panels of the storyboards are scanned into a computer to give a 'rough cut' of the show
- After a number of the adjustments, the detailed animation built to produce moving colour footage with voices
- Final draft viewed frame-by-frame
- List of retakes made. To be shot and dropped in later
- Final film edited to required length.

You can see from the above that the making of an animated film is a long and complex expensive business requiring a large number of people working together as a team.

The process is similar to feature film making, in that the whole effort is like a military operation with very clearly defined job descriptions. The main difference is in the sequence of events.

Training

Training in the animation industry is at the top of the agenda for many industry specialists. Recruiting and developing people who can apply a mix of craft, technical, interpersonal and business skills is a difficult task. Some of the UK's leading companies have become involved in the provision of relevant, practical training. The Bristol Animation Course is perhaps the best example of this, supported by a group of industry specialists, who teach on the courses and also provide supervised work experience. Its aim is to develop talented but 'raw' animators. Students work an eight-hour day, five days a week and quickly learn about the demands of the industry. Another course, which has evolved for similar reasons, is the London Animation Studio, based at the St Martins School of Art in London. This course concentrates on drawn and computerised animation, and in six months aims to develop industry standard skills which should make individuals more marketable. It is open to a diversity of people including those who already work in the industry.

During the Bristol course, students are assessed on a variety of attributes, including:

- Observation/characterisation
- Movement/ animation
- Staging/editing/storyboarding
- Computer skills

- Attitude to work

Employers are keen to support training that develops a range of skills needed by the industry. Skillset professional qualifications have been designed to incorporate these and a number of assessment centres have been established. Attributes sought after in these centres include:

- The ability to communicate ideas clearly
- Persuasiveness
- Patience
- Attention to detail
- Lateral thinking
- Ability to work under pressure
- Numeracy/financial ability
- Fundraising/talking to investors

Areas of particular skills shortages have been identified for scriptwriters, producers, storyboarders and layout artists.

Other training and development opportunities of interest include the following:

The British animation training scheme

Established in the mid 1990's, funded by skillset, CARTOON and the industry, 12 people working within the sector are selected each year to attend on a day release basis.

Finding out more

The Animation UK Directory – published annually by Venture publishing, in association with BECTU, contains information about:

- Specialist production companies, including special effects, multi media, games designers etc;

- People who work in all areas of the industry with names and contact details;
- Relevant service providers, including equipment hire and sales, specialist consultants and agents
- European production companies
- Other useful information

Animation UK also publishes an industry magazine. Visit their website www.animationuk.com.

9

THE FACILITIES SECTOR

One other area directly connected to television is that of the facilities sector. More and more people are choosing to start their career in film and television in this sector.

The facilities sector supplies a wide range of specialised technical services to a range of media companies. This sector is very well established in the United Kingdom and is very competitive. It is a fact that almost every new production will call on the services of one or more aspects of the facilities sector, including broadcast television, commercials, feature films, promotional videos, studios, the corporate sector and so on. Although the sector is competitive and much work goes to television and commercials the film industry generally calls upon the services of the sector.

Facilities companies supply all types of technical backup, such as camera equipment and crews, film-processing laboratories, edit suites and editors and also audio services. The sector has grown in response to the growth of independent production companies and deregulation of large broadcasters. With expanding need there has developed a supply sector. The main advantage of utilising the facilities sector is that the companies offer trained staff and technology for short-term use, thereby overcoming the obstacle of expensive employment costs.

Opportunities in the facilities sector

Recruiters are particularly keen to attract people with well-developed interpersonal skills. Customer services skills are in demand at every level. If an individual also has a high level of technical expertise then this is a distinct advantage.

Although there are some big players within the facilities sector, including BBC Resources, The industry is largely made up of small specialist companies. The number of employees will vary depending on the size of the company.

Skills and personality factors required

Many people join facilities companies at the start of their careers and develop the relevant technical skills on the job, so to speak, sometimes supported by specialised off-site training. At early recruitment stages, managers will be looking for:

- Proven technical ability
- Knowledge of IT
- Awareness of the commercial environment
- An open, customer friendly approach
- Good time management skills
- Flexibility about working hours.
- Good general business skills
- A genuine enthusiasm for the industry

The industry is in permanent need of well-trained and motivated staff and there are a variety of jobs on offer:

- On-line editors
- Off-line editors
- Telecine operators
- “D Graphics animator
- 3D Graphics animator
- Graphics designer
- Audio engineers
- Audio assistants
- VT Operator
- Quality operator
- Transmission engineer

- Camera operator
- Studio VT assistant
- Vision mixer
- Lighting supervisor
- Sound recordist
- Engineer
- Electrician
- Operations manager
- Facilities manager
- Studio manager

There are also jobs on the business side, including:

- Contracts administration
- Bookings assistant
- Sales and marketing director
- Managing director
- Reception
- Financial manager
- Business development executive

The main entry job within the facilities sector is that of 'Runner'. As a runner you will get many opportunities to show off your potential skills and also your enthusiasm. In an independent facilities house, you could expect the following type of career progression, if you have talent and personal skills: Runner – VT Operator – Editor's assistant – Junior editor. (2yrs)

Editors can earn a large amount of money, as can other jobs within the industry. The facilities sector is a very good place to commence a career in film generally.

Organising some work experience where you can get a feel for the industry would be invaluable, and gaining experiences in several different companies would help more. The companies, together with their specialisms, are listed in trade directories, such as The Production

Guide Kays UK production Manual, Kemps Film, TV and Video Handbook, and The Knowledge. Once you have decided on the filed you wish to specialise in you should apply to the relevant company, sending a C.V emphasising your strong points.

Having a degree would certainly be useful. This may help you and give you a better chance than those without degrees. Students generally gain interpersonal skills during their studies and are often considered to be stronger candidates.

The majority of facilities companies can be found within the Soho area of London, although there are a number of the larger ones spread around Greater London. Around 25% of the companies are in other regions of the U.K.

Training and education

Some colleges and universities have well-established links with employers and are careful to ensure that their syllabus includes many marketable elements. By contacting the employers and the colleges that you are interested in, you should be able to develop a shortlist of possible courses. Full time courses are listed in Media Courses (2002) and a wide range of short courses in Media and Multimedia Short Courses. Both are published by the British Film Institute and should be available in libraries. Skillset Professional Qualifications will be relevant to many jobs in this sector. Read chapters 9 and 10 for more information about courses.

Finding out more

Televisual – published by Centaur Communications is a trade magazine read by the facilities industry. In September of each year it publishes an overview of the industry, nominating the best companies. *Broadcast,* published by EMAP, has a dedicated facilities section in each magazine and useful occasional specialist supplements. *The Production Show,* held in London each year (usually in March), is the best showcase for current available technology with many relevant companies exhibiting.

10

CAREERS IN TELEVISION

No doubt you are exactly the kind of person that the television industry has been waiting for (a little self-confidence will get you through some of the rejections you are going to have to live with!) but people on the inside are kind of old-fashioned, they hear you talk the proverbial talk, but they will expect you to back it up with your proven skills, and nothing impresses quite as much as good old-fashioned qualifications.

You can train for a variety of careers, in a variety of places, but knowing what to look for, and where to find it, will save you time, trouble, and money, so take the time to read on, you'll be glad you did. Media Studies courses are very very popular – everyone sees themselves as a TV star, so everyone figures a few years preening and pouting at their local university or college will get them the boring but necessary slip of paper to head straight to the front of the queue. This, and you being a television realist will know, is utter nonsense. Media Studies courses were seen as a glamorous option for students who had some sort of woolly idea about being the new Melvin Bragg, and they brought in lots of money to the universities and colleges that offered them, so it wasn't long before any academic institution was offering a Media Studies course to anyone who could front the cash and could hold a clipboard and talk loudly about the psychology of car adverts in the Union bar. The rapid rise of the Media Studies course meant that far from every course was valid, or relevant, or even worthwhile, and before long the media got hold of the theory that a Media Studies course ranked slightly below staring out of the window composing epitaphs to obscure European poets, and slightly above just staring out of the window, as a valid way for a student to spend his or her time in the rarefied world of study.

Television hasn't got where it is today by employing windbags with qualifications that fit them for the world of after-dinner conversation and precious little else, and it realised that taking a hand in the development of appropriate education would ensure that it could rely on a steady stream of properly qualified people to actually work in its industry – and as you will have gathered by now, work in television means just that.

First of all you should be sure that the course you are looking at is the one you need for the job you want to do. An excellent start is understanding the difference between 'Media Studies' and 'Media Production'. The first is an academic qualification that will enable you to write and speak with authority about the media – the social impact of 'reality television', and the worth of soap operas as social barometers, which is all well and good if that's what you plan to do with your life.

If you are reading this book, then you need to be aiming for a 'Media Production' course which will give you solid hands on experience of how television programmes are made, and how your skills can best be utilised in making them, which is obviously more likely to suit your chosen career aspirations.

So, now you know what kind of course you want to find – the next step is where to find it.

Your university of choice should be selected on the basis that it offers you the Media Production course best suited to your needs. You should enquire if the university has links with the local television company – this may be your first employment opportunity, so you should choose your university carefully – base your opinions on the surrounding living and working environment, and not just the wide range of obscure beers carried by the Union bar – this place could be your home and workplace for a few years, so try and find somewhere you are going to like, it will make the sheer slog of getting your foot in the television door that little bit easier to deal with.

Check that you get a decent amount of work experience on your course, it's all very well knowing all about lenses and lighting, but if you've never stood behind a camera before, the director won't be too keen to give you chance to work out what all the buttons and knobs are for, remember, time is money, in television that maxim applies every minute of every day, don't be the one who wastes either of them.

Make sure that your course is geared towards the area in which you want to work – you should know if you are better suited to the technical side, production, front-of-camera and so on, and a good course will be one that will give you proper training and experience for your chosen work area.

It might be worth checking a few contacts in your proposed field of employment and see if anyone can recommend good training courses – television people usually remember how they got started, and hopefully they'll be willing to pass on some of the knowledge to you, so ask around, and write and phone the appropriate contacts.

You should be aware, but it's worth pointing out anyway – television is a highly competitive work arena, you've got to be good at what you do, a team player, willing to listen, learn, and adapt to changing scenarios – before you actually arrive for your first job. Train hard, train right, and you will have your foot on the bottom of the greasy pole that has a television career at the top. Will it be worth it? If you need to ask, you're reading the wrong book!

Formal training for any occupation shows a level of commitment that is attractive to prospective employers, but it is no guarantee of future employment in television.

You may find that having achieved paper qualifications, you are in a queue behind people who have joined a television company from school in secretarial jobs, or as runners, who are already on the inside track. That should not deter you from considering training courses – the sheer proliferation of available training opportunities proves a large demand for qualified personnel, particularly within the technical areas of television production.

In the 'bad old days' of television, would-be television industry employees were thwarted by a 'Catch 22' situation which resulted in all manner of devious methods of circumvention. Specifically, you were unable to obtain paid employment in television, unless you could produce proof of membership of the appropriate Trades Union – membership that could only be obtained by, surprise surprise, proof of paid employment!

Happily, those days are now gone, and entry into television merely relies on talent, luck, a large dollop of 'right place right time', and similar variables of which you should be well aware having studied the previous chapters of this book.

Anyone who has wondered about finding a suitable training course over the last ten years has had the unbelievable good fortune of being able to take advice from SKILLSET (The Sector Skills Council For The Broadcast, Film, Video and Interactive Media) which will give you a wonderful starting point in the bewildering world of television employment. Their excellent (and you would expect it to be!) website will guide you into the specialist areas you need, and there is a section on related websites for you to explore further options.

If you are still unsure about your career options, it is well worth considering a 'taster' workshop or Masterclass of the type run by the Royal Television Society, contact them at Holborn Hall, 100 Grays Inn Road, London WC1 8AL Telephone 020 7430 1000. E-mail ifo@arts.org.uk.

As the oldest Broadcasting Company, The BBC has the longest established, and in many opinions, the best training facilities available in television.

Naturally with such excellence comes a training facility that is vastly over subscribed. The BBC's trained staff are regarded as the finest available, and you can expect to have to be well above average to not only complete one of the training courses, but to be accepted for entry in the first place. If circumstances dictate that you need to train in your own locality, it may be possible to pick up a place on a course near to your home.

Technically minded trainees can apply to join as Engineers. You will need to be at least eighteen, and GCSE's Grade C or above in English and Maths with Physics, as well as Maths with Physics at 'A' Level, although a BTEC Diploma in Electrical / Electronic Engineering is acceptable in lieu of the 'A' Level requirement. For details, contact – BBC Recruitment Services PO Box 7000 London W1A 6GJ Telephone 020 7765 1192 E-mail recruitment@bbc.co.uk

For construction, production and technical staff, the excellent **FT2** (Film And Television Freelance Training) provides an excellent training programme for proposed freelance technical staff within the television industry. The training takes an "apprenticeship" format, with practical placement within various television production areas, as well as in-house training provision. The courses lead to qualification in the Skillset NVQ's at Levels 2 and 3.

Selection for these courses is rigorous, and as you would expect for a training system which is both thorough and Industry recognised, and there are more than sixty applications for each place. Applicants will have to demonstrate an appreciable level of interest and commitment to television, and any appropriate qualifications which can be attained in advance of application will assist in the final selection procures.

For those employed in television (i.e. those who have gained a foothold by signing on as secretaries, runners, and so on) who wish to obtain training for their proposed specialist areas, there is an

11
THE FUTURE OF TELEVISION

It all comes down to money. Not art, or culture, or information, or entertainment, just money.

That may sound cynical, but television is a business – it costs thousands to produce its output, but pence to actually watch and enjoy it, and that system has embraced the production of television almost since it started. The 'money' aspect comes from a need to procure in order to produce television, and hopefully some revenue as a result, although sources of either can vary considerably.

The British television viewer has lived with the payment of a Licence Fee for generations – it's grudgingly accepted as 'the way things are', although it may not be so for much longer, but we'll come to that in a moment. The simple premise of effectively paying nothing to watch half the service, provided by Commercial Television, whilst paying an annual fee to watch the BBC whether viewers want to or not has been the norm for as long as anyone can remember.

As the new century came into being, the increasing viewpoint has been that people should actually pay for what they want to watch, and thus 'market forces' will dictate viewing habits. This sounds like an alarming concept, until you realise that print media has survived in this marketplace since its inception, and no one appears to mind paying for what they want, and newspapers have been made to adapt to audience trends in order to survive.

No need to panic just yet – the current situation seems to suit the main players just fine.

The BBC can adhere to its remit to 'educate inform and explain' which allows it to present a variety of dramas, documentaries and entertainment that is valued and envied throughout the world. The commercial stations can woo its advertising lifeblood and plough a wide, if comparably shallow furrow of populist entertainment, and the satellite stations can encourage people to pay for a raft of output that they may not want, in order to obtain the output that it actually signed up for in the first place. It suits almost everyone, but it may not stay that way forever.

Television has always evolved, and will continue so to do – the advent of digital television has speeded that evolution from a slow and graceful curve to a straight-up line with scorch marks on it. From now on, the endless choice provided by the ever expanding digital network will mean that more and more channels are competing for, in relative terms at least, fewer viewers. The survivors will be those that can provide steady advertising revenue, or those that can use airtime to sell the channel proprietor's products. A possible future for television in the UK can be observed by taking a look at our European neighbours.

In France, The M6 channel has taken a mere fifteen years to evolve into the second largest commercial broadcaster in the country. It has done so by identifying its audience, and directing its format and advertising accordingly. The station's demographic is the high-spending mid-teens to mid-thirties, and it works by directing a significant proportion of its advertising towards the parent company's products which include video and CD outlets, events, magazines, and so on. It hasn't taken long for other companies to see the strength of M6's advertising campaigns for its own divisions, and to add their own advertising revenue to its rapidly expanding bank balance. This type of programming and advertising could well be a future role model for the digital age in the UK

As television makers realise the increasing demand for revenue-driven programming, output will adapt accordingly. Examples of this type of approach can be seen in the success of programmes like

'Big Brother', an outstanding success throughout the European audiences, it proved a similar success here with its profits being created by the culture of 'phone voting', which effortlessly raises serious amounts of money without the need for expensive production values. Once the initial setting – in this case a custom-built house erected relatively inexpensively in a 'non des-res' area, was completed, the actual transmission costs were relatively low, with the 'cast' coming free of any charge, except perhaps to their dignity! That has to be television making at its zenith, which is not to say that this is the way television should go, but it may well form the basis for a large proportion of future viewing habits.

So how does this affect the prospective employee in the television industry? Prospects continue to improve. Because there is an increasing number of stations springing into being, there has to be increased output to fill their airtime, and this means more requirements for every skill that television as a medium requires. Technically, the digital age will see an increase in the studio technology, and prospective staff will be advised to ensure they keep abreast of the developments as far as possible – the employment opportunities will doubtless expand, but that doesn't mean that the queue of people waiting to fill them will necessarily diminish!

Television has never been a more exciting or challenging place in which to work than it is today. It will be for the technicians, front-of-camera people, companies and individuals to shape its future – if you've read this far, that thought should excite you beyond measure. It's out there, all you have to do is go and get it.

AND FINALLY .....

The world of television is a vibrant and exciting place to work. The pressures are considerable, and in terms of satisfaction, so are the rewards. Television is a world that appeals to vast numbers of people, some, but not all, who will have a distinctly rosy-coloured view of what working in television is all about. If you have read all the chapters in this book, then you will not be included in that happy band, and you will have decided if the world of television production is somewhere you can be happy. Unless it is a job you want more than any other, and you are prepared to overcome all the many and various obstacles, then you should move over and let your prospective place be taken by one of the other thousands of people who know that this is the career they have always wanted.

Television is a vast and complex structure, and the vast majority of people who spend their working lives creating it for the education and amusement of audiences around the world are its unsung heroes. Technical staffs have to be the very best in their particular disciplines. Television costs pence to watch, but thousands to produce, and the majority of the money is eaten up by the complex technology used to make television programmes. Because there often isn't the time, or the money, to re-shoot scenes several times to get them right, you can be expected to work on a 'first time every time' basis. If you don't respond well to working under pressure, and producing miracles on a daily basis, only to find that Directors and Producers accept your efforts as no more than you doing your job, then you should be looking at life in industry instead. Get your training, get your experience, and be the best you can every day, because the competition is always out there, and the weakest go to the wall.

You may have to begin your television career in an area other than that for which you are trained, and where your skills will be best utilised. For that reason, keep your options open – be ready to accept a job as a Runner or a Secretary just to get yourself on the inside of a television company. There are various media vacancy outlets, but there is no substitute for being on the inside, and learning exactly where the power lies, what individuals actually want and need from the teams they build around themselves, and being able to pitch yourself for a job before the outside world even knows it is available.

Making a good impression is vital if you are to get a head start in the highly competitive job market. This means you must get your CV prepared professionally, find out how to write an appropriate business letter, and make sure you tailor it to the individual to whom you are writing – form letters are impolite, and show a basic lack of understanding of the way television works. Anyone in a position to employ you as a new starter has worked long and hard to get where they are – the appropriate level of respect and knowledge based on whom they are, and what they do, will always set you in good stead, and ensure that your letter doesn't hit the bin after two lines of the 'Dear Sir or Madam, any chance of a job' approach!

Always know your skills and strengths and present them in a reasonable manner – you will be called to walk the walk if you manage to talk the talk, so it's pretty futile trying to make out you are something you're not. By the same token, don't be a shrinking violet – you are out to impress someone that you are worth employing. Don't fall for the British habit of being over-modest – if you convince people that you may be up to the job, and you'd like to have a go, they will simply reject you in favour of someone who is prepared to state what they can deliver.

Remember that rejection is a natural part of the industry selection process – it's not personal, and it doesn't mean you are not good enough. It just means you weren't the right person on the right day. Always endeavour to turn a rejection into a positive asset – try and get some feedback from the interviewers about your technique at interview, your CV, your qualifications, and so on, and use their advice to good effect for the next time.

Television is at the dawn of a new and exciting period in its history – there has never been a better time to join in at the beginning. The advent of digital stations, and the explosion in television audience levels means there has never been a greater demand for television programmes to fill the ever expanding station schedules, and that means a corresponding increase in the people who can make those television programmes happen. If you have taken notice of the advice given in the previous pages, you should be equipped to avoid the obvious pitfalls in getting a job, and you should have obtained the necessary training to perform your tasks well, and ensure a reasonable continuity of employment, not forgetting that the trend in employment is increasingly towards short-term contracts.

Television programmes are out there waiting to be made. All you have to do is go out there and make them!

12

FURTHER INFORMATION

So far, we have looked at the television industry, including the animation sector plus commercials and documentaries and have outlined jobs within these areas. Outlined below are some useful trade associations and publications through which you may gain useful information that will help you find out more about the industry and specialised areas. In addition, in the next chapter there is a list of short and long courses should you wish to take a course of further education.

Trade associations, guilds, unions, societies and support organisations.

The media industry generally is supported by a broad network of associations, trade unions and guilds, which act on behalf of their membership. A selection of the most useful is listed below.

Advertising, Film and Videotape Producers Association
28 Noel Street
London W1V 3RD
020 7434 2651
www.afvpa.com

This organisation represents the interests of producers of commercials, and has negotiated agreements with the relevant unions. The AVFPA is also a member of the commercial film producers of Europe.

Amalgamated Engineering and Electrical Union (AEEU)
Hayes Court
West Common Road
Bromley
Kent BR2 7AU
www.aeeu.org.uk

The AEEU is a large multi trades union with approximately 1000 members involved in film and TV production.

British Film Commission (BFC)
10 Little Portland Street
London W1W7JF
www.britfilcom.co.uk

Now part of the Film Council the BFC is supported by government funding, the BFC promotes the UK as an international production centre and provides support for those filming in the UK.

British Film Institute
21 Stephen Street
London W1P 1PL
020 7255 1444
www.bfi.org.uk

The national agency for conserving film and broadcast. The BFI has a library open to the public and also offers research.

Broadcasting, Entertainment, Cinematograph and Theatre Union (BECTU) 111 Wardour Street London W1V 4AY 020 7437 8506
www.bectu.org.uk

BECTU covers all technical production staff in broadcasting, film and independent production. It has agreements for these sectors and offers members representation, employment, legal and taxation services. It publishes a regular journal, 'Stage, Screen and Radio' and runs a student link up scheme for people intending to build a career in the industry. Membership provides a range of opportunities to establish contact in the industry and to gain up to date information from the journal. Students researching relevant projects will also have access to relevant advice and help from BECTU. Students and new graduates are offered discounted membership rates.

The BECTU Black Members Sub-Committee was set up to help and support members of Afro-Caribbean origin or anyone who may suffer discrimination in the workplace.

British Academy of Film and Television Arts (BAFTA)
195 Piccadilly
London W1V OLN
www.bafta.org

Bafta promotes high standards in film and TV production and has an annual awards ceremony. It holds extensive programmes of lectures, seminars etc.

British Interactive Multimedia Association (BIMA)
5-6 Clipstone Street
London W1P 7EB
020 7436 8250
Email enquiries@bima.co.uk

The trade body for the multimedia industry, it publishes a directory of members, regular newsletters and arranges a series of industry meetings.

The Moving Image Society (BKSTS) 63-71 Victoria House, Vernon Place, London WC1B 4DA 020 7242 8400 www.bksts.demon.co.uk

This organisation organises regular meetings and demonstrations of new equipment and techniques and runs training courses. BKSTS also organises a biennial conference and specialist seminars.

British Screen Finance
10 Little Portland Street
London W1W 7JF
Email info@britishscreen.co.uk

British Society of Cinematographers (BSC)
11 Croft Road
Chalfont St Peter
Gerrards Cross
Bucks SL9 9AE
O1753 888052
Email Britcinematographer@compuserve.com

The BSC arranges meetings, film shows etc.

Broadcasters Audience Research Board (BARB)
Glenthorne House
Hammersmith Grove
London W6 OND
020 8471 9110

BARB produces statistical information which is used to calculate the ratings.

Broadcasting Standards Commission (BSC)
7 The Sanctuary London SW1P 3JS 020 7223 0544

The BSC is the statutory body for both standards and fairness in broadcasting.

Community Media Association
15 Patemoster Row
Sheffield
S1 2BX
0114 279 5219
www.commedia.org.uk

Provides training and holds conferences and events.

CSV Media
237 Bentonville Road
London N1 9NJ
020 7278 6601
Email information@csv.org.uk

Part of the national charity Community Service Volunteers, CSV Media specialises in social action broadcasting, media support services and media training.

Directors Guild of Great Britain
Acorn House
314-320 Grays Inn Road
London WC1X 8DP
020 7278 4343
www.dggb.co.uk

Represents the interests and concerns of director in all media. It seeks to maintain high standards in the film, television and theatrical media.

Film Council
10 Little Portland Street
London W1W 7JF
www.filmcouncil.org.uk

The Film Council plays a leading role in leading the development of the British Film Industry.

Guild of British Animation
26 Noel Street
London W1V 3RD
020 7434 2651
Email afvpa@easynet.co.uk

Represents the interest of British animation companies

Guild of British Camera Technicians
5/11 Taunton Road
Metropolitan Centre
Greenford
Middlesex
UB6 8UQ
020 8578 9243

The main aim of the Guild is to represent and further the professional interests of technicians working with motion picture cameras.

Guild of British Film Editors
Travair, Spurlands End Road, Great Kingshill, High Wycombe, Bucks HP15 6HY. 01494 712313.
Aims to increase the recognition of film and sound editing as part of the creative and artistic aspects of film production.

International Visual Communications Association (IVCA)
Bolsover House
5/6 Clipstone Street
London W1P 8LD
020 7580 0962
www.ivca.org

IVCA is the professional trade association representing the interests and needs of the visual communications user or supplier. It has over 1500 members representing programme producers, commissioners, users, in-house units, facilities, manufacturers and others.

Music, Film and Video Producers Association (MFVPA)
26 Noel Street
London W1V 3RD
020 7434 2651

The MFVPA represents the major music video companies. It has an agreement with BECTU.

PACT (Producers Alliance for Cinema and Television)
45 Mortimer Street
London W1N 7TD
020 7331 6000
www.pact.co.uk

PACT serves the film and independent TV production sector and is the UK contact point for international co-production or co-financial partners and distributors. The monthly PACT magazine is useful as a source of information about the independent sector. PACT also runs an industrial relations service which provides information on standard agreements and contracts, and negotiates with the relevant unions.

Production Managers Association (PMA)
Ealing Studios
Ealing Green
London E5 5EP
020 8758 86999
www.pma.org.uk

Provides training job opportunities and support for experienced production managers.

Scottish Screen
249 West George Street
Glasgow
G2 4QE
0141 302 1700
www.scottishscreen.demon.co.uk

A government backed body encouraging film development and education in Scotland.

Sgrin-Media Agency for Wales
The Bank
10 Mount Square
Cardiff Bay
Cardiff CF1 6EE
029 2033 3304
www.sgrinwales.demon.co.uk
The Welsh national information centre for European audio-visual funding and policy in Wales. It promotes the EU action programme for the audio-visual industry.

Skillset
103 Dean Street
London W1V 5RA
020 7534 5300
www.skillset.org

The national training organisation for TV, radio, film, video and interactive media. Recognised by the government as the voice of the industry in training. Skillset operates at a strategic level to improve training and education policy on provision.

Women in Film and Television
6 Langley Street
London WC2H 9JA
020 7240 4875
Email wftv@org.uk

Professional membership organisation for women working in the film and TV industries. Has an annual awards ceremony and campaigns on behalf of its members.

The Writers Guild of Great Britain
430 Edgeware Road
London W2 1EH
020 7723 8074
Email tostia@wggh.demon.co.uk

The writers union, providing support and training for its members.

Specialist Film Industry Organisations

In addition to the above, there are a number of other specialist organisations associated with the film industry. The major players are listed below.

The Production Guild
C/O Sargent-Disc Ltd
Pinewood Studios
Pinewood Road
Iver Heath
Bucks
SL0 ONH
01753 657167

Guild of Location managers
C/O The London Film Commission
20 Guston centre
Regents Place
London NW1 3JH
020 7387 8787

British Special Effects Guild
Box 23 Shepperton Studios
Studios Road
Shepperton
MiddlesexTW17 OQD
01932 572631

Association of Motion Picture Sound
28 Knox street
London W1H 1FS
020 7402 5429

British Film Designers Guild
9 Elgin Mews South
London W1P 7PJ
020 7286 6716

First Film Foundation
9 Bourlet Close
London W1P 7PJ
020 7580 2111

(A charity which exists to help new British writers/producers and directors based in the UK and Ireland, to make their first film. Offers impartial practical advice and several programmes for training and development)

New Producers Alliance
9 Bourlet Close
London WJP 7PJ
020 7580 2480

(A membership organisation for producers, directors and scriptwriters which produces a directory of members and suppliers)

Film Education
020 7976 2291
www.filmeducation.org

(offers cross-curricular programmes designed to improve understanding of the film industry)

The Short Film Bureau
01273 2355525
www.shortfilmbureau.com
(Offers help to assist film-makers to get their work shown commercially)

British Council Film, TV and Video Department
11 Portland Place, London W1N 4EJ. 020 7389 3065

Film Four
76-78 Charlotte Street
London W1P 1LX

London Film and Video Development Agency
114 Whitfield Street
London W1P 5RW
020 7383 755

NESTA
33 Throgmorton Street
London EC2N 2BR

The Northern Ireland Film Commission
21 Ormeau Avenue
Belfast
BT2 8HD
02890 232444
www.nifc.co.uk

Useful Publications

Magazines and journals

Many of the publications listed below can be found in specialist libraries. Alternatively, the larger newsagents will order them on your behalf.

Animation UK
News about the growing animation industry

Ariel
The BBC in-house magazine

AW Magazine
Up-to-date information non the audio-visual business.
Direct
The newsletter of the Directors Guild of Great Britain.

Eyepiece
Introduced by the Guild of British Camera Technicians, it is also full of relevant information for aspiring film makers

Image Technology
Magazine of the Moving Image Society. Useful for anyone planning a technical career in film.

Moving Picture International
Provides an insight into the international film, TV and Video Industry. Only published during festivals.

Production Solutions
For people working in TV and film production, with an emphasis on technology, film issues and training.

Screen International
Invaluable for aspiring film-makers, it contains interviews and features.

Reference Books

The books listed below are a selection of useful reference works. They can be obtained from bookshops or specialist or business libraries. Check with publishers for prices.

Animation UK (Venue Publishing)
64-65 North Road, St Andrews, Bristol BS6 5AW. 0117 942 8491
The directory of the animation industry

BFI Handbook (BFI)
21 Stephen Street
London W1P 2LN
020 7255 1447

Contains facts and figures and addresses and contacts within the industry.

The Creative Handbook (Variety Media Publications)
34-35 Newman Street
London W1P 7FB
020 7436 8626

A-Z of Britain's Film, TV, Radio and Theatre Directors.

Directors Guild Directory (Directors Guild of Great Britain)
15-19 Great Titchfield Street
London W1P 7FB
020 7436 8626

Directors and Producers Directory (BECTU)
111 Wardour Street
London W1V 4AY
020 7437 8506

Directory of British Film and TV Producers (PACT)
45 Mortimer Street
London W1N 7TD
020 7331 6000

Directory of International Film and Video Festivals (British Council) 11 Portland Place, London W1N 4EJ. 020 7389 3065

The IVCA Business Media Handbook (IVCA)
Bolsover House
5/6 Clipstone Street
London W1 8LD
020 7580 0962

The essential guide to the corporate visual communications industry, providing facts, figures and contacts.

Kays UK Production Manual (KAYS)
8 Golden Square
London W1R 3AF
020 8749 1214

A comprehensive manual of people and organisations in the production side of film. Contains thousands of names and addresses.

Kemps Film, TV and Video Yearbook (Variety Media Publications)
34-35 Newman Street
London W1P 3PD
020 7637 3663
Long established directory of the film and television production industries in most countries.

The Knowledge (Benn Business Information Services)
Riverbank House
Angel Lane
Tonbridge
Kent TN9 1SE
01732 362666
A comprehensive guide to the products and services of the UK film, TV and Video industry.

The Media Guide (Fourth Estate)
020 7727 8993

A source of reference for anyone planning to work in the industry.

The Production Guide (Emap Media)
33-39 Bowling Green Lane
London EC1R ODA
020 7505 8000

Annual details of technical contacts, services and equipment.

The White Book
01932 572622
International production directory

The Art of the Deal
Dorothy Viljoen PACT
Comprehensive guide to business affairs for film and TV producers.

13

GUIDE TO COURSES AND COLLEGES OFFERING COURSES

It is important before you undertake a course to make sure you find out whether it has some legitimacy and carries some weight with people in the industry. It is also very important to make sure that you choose a course that is relevant to what you want to do after you have finished. It is no good, for example, choosing a highly theoretical course in film history if you intend to embark on a career in sound or lighting. It is always necessary to receive the right advice before you commit yourself.

The courses listed below are all further education courses and short courses and are highly practical. There are many courses that are theoretical but you will need to obtain university and higher education handbooks to determine the whereabouts of such courses.

For up to date information concerning the courses listed below you will need to contact the school concerned.

The courses are normally two years long and will often incorporate an industrial placement. The entry requirements are usually 4 GCSE's grades A-C, a BTEC or SCOTVEC National Diploma or Certificate or their equivalent. As with many courses, exceptions are made for mature students with no formal qualifications.

The first section below lists a cross section of specialist film related subjects, with the appropriate college listed underneath. For contact details of the college concerned you should look in section two which lists a broad cross section of colleges and universities in the United Kingdom offering film courses. However, given that there are also many undergraduate and post-graduate courses on offer the colleges listed below represent only a small cross section of further education courses available. For a comprehensive guide to all media courses, 'Media Courses UK' can be purchased from the British Film Institute. Price: £13.99. Alternatively, this information is now on line at the British Film Institute web site.

(1) Specific courses in television and film related disciplines

ACCOUNTANCY

BA (Hons) Accountancy and Film Studies (Modular)
Bolton Institute of Higher Education

ANIMATION

Animation
Westminster Kingsway College

Animation and Multimedia
Tower Hamlets College

Animation Assistant Course
British Animation Training Scheme

BA (Hons) Animation
Brighton College of Technology

BA and BSc (Hons) Animation
University of Glamorgan

BA (Hons) Animation
University of Lincolnshire and Humberside

Surrey Institute of Arts and Design
University of Wales Newport
University of Wolverhampton
BA (Hons) Animation and Illiustration
Southampton Institute

BA (Hons) Animation, Media and Society
Northbrook College

BA (Hons) Another subject with digital animation
London College of Music and Media

BA (Hons) Computer Animation
University of Teeside

BA (Hons) Computer Visualisation and Animation
Bournemouth Media School

BA (Hons) Design Animation
North East Wales Institute of Higher Education

BA (Hons) Design: Animation
Staffordshire University

BA (Hons) Design for Interactive Media with digital animation
London College of Music and Media

BA (Hons) Film and Animation Production
The Arts Institute at Bournemouth

BA (Hons) Graphic Arts Animation
Liverpool John Moores University

BA (Hons) Graphic Design (Animation)

Norwich School of Graphic Art and Design

BA (Hons) Media Production with Animation
University of Luton

BA (Hons) Multimedia Design and Digital Animation
Cumbria College of Art and Design

BA (Hons) Photography and Digital Imaging with Digital Animation
London College of Music and Media

BA (Hons) Visual Communication (Animation)
Edinburgh College of Art

BDes (Hons) Animation and Electronic Media
Duncan of Jordanstone College of Art and Design

BSc Computer Animation and Special Effects
University of Bradford

BSc (Foundation Degree) Computer Visualisation and Animation
Ravensbourne College of Design and Communication

BSc (Hons) Multimedia Computing with Digital Animation
London College of Music and Media

HNC in 3D Computer Animation
James Watt College

HNC in Filmcraft and Animation
Motherwell College

HND in 3D Computer Animation
James Watt College

HND in Animation
Glamorgan Centre for Art and Design Technology

HND in Animation and Creative Media
Plymouth College of Art and Design

HND in Animation Production
James Watt College

HND in Design Technology (Multimedia, Broadcast Graphics and Animation)
Farnborough College of Technology

HND in Filmcraft and Animation
Motherwell College

LOCN Animation and Multimedia
Tower Hamlets College

MA in Animation
North East Wales Institute of Higher Education
Royal College of Art

MA in Film and Television (Animation Direction)
National Film and Television School

Mphil and PhD in Animation
Royal College of Art

National Diploma in Media and Animation with Model making for Film and Television Bridgewater College

WNYOCN Animation Workshop – Level 3
York College

BUSINESS

BA (Hons) Advertising with Business
London College of Music and Media

BA (Hons) Business Economics and Film Studies (Modular)
Bolton Institute of Higher Education

BA (Hons) Business Information Systems and Film Studies (Modular)
Bolton Institute of Higher Education

BA (Hons) Business Studies and Film Studies
Bolton Institute of Higher Education

BA (Hons) Communicatioon Studies (EFL) and Business
Buckingham University

HND in Business (Arts and Media Management)
New College Durham

CAMERA

Camera Operations – Max Q
Line Out

NVQ Camera (Level 2)
Grimsby College 35

NVQ Camera (Level 3)
Grimsby College

CINEMATOGRAPHY

MA in Film and Television (Cinematography – Film and Video)
National Film and Television School

COMPOSING

Bmus (Hons) Performance and Composition
London College of Music and Media

Composing and Synchronising Music to Film and Video
Westminster Adult Education Service
MA in Composing for Film and Television
Kingston University

Mmus in Composing for Film and Television
London College of Music and Media

DIRECTING

Directing Actors for Camera
North Kensington Video-Drama Project

Directing Diploma
ARTS International

Documentary Production and Direction
North Kensington Video – Drama Project

MA in Film and Television (Animation Direction)
National Film and Television School

MA in Film and Television (Documentary Direction)
National Film and Television School

MA in Film and Television (Fiction Direction)
National Film and Television School

ECONOMICS

BA (Hons) Business Economics and Film Studies (Modular)
Bolton Institute of Higher Education

BA (Hons) Media and Popular Culture with Economics
University College, Northampton

EDITING

BA (Hons) Editorial and Advertising Photography
Kent Institute of Art and Design

Editing – Non-linear and Linear – Max Q
Line Out

HNC in Broadcast Post production
Ravensbourne College of Design and Communication

MA in Film and Television Editing
National Film and Television School

NVQ Edit (Level 2)
Grimsby College

NVQ Edit (Level 3)
Grimsby College

PgCert and PgDip in Script Editing
Liverpool John Moores University

HAIRDRESSING

Film and Television Make-Up, Hair Design, Wigmaking and Special Effects Course
Retford International College

Historic Hairdressing NVQ Level 3 (Long Hair, Wig and Hairpiece Design Attachment)
Retford International College

Make-Up and Hairdressing – Toatl Look Fashion
Retford International College

The Complete Make-up Artist Training programme
Delamar Academy of Makeup

MAKE-UP

Certificate of Higher Education in Professional Make-Up Design
Delamar Academy of Make-up

Edexcel Professional Development Certificate in Media Make-Up
Westminster Adult Education Service

Film and Television Make-Up, Hair Design, Wig making and Special Effects Course
Retford International College

HND in Theatrical and Media Make-Up
Somerset College of Arts and Technology

Make-Up and Hairdressing – Total Look Fashion

Retford International College

Make-Up Artists Diploma
North-east Worcestershire College

National Diploma in Media Make-Up
Stockport College in Further and Higher Education

Speciality make-Up
Retford International College

Television, Film and Video Make-Up
Retford International College

The Complete Make-Up Artist Training programme
Delamar Academy of make-Up

Theatrical make-Up (Character)
Retford International College

Total Look – Advanced
Retford International College

Total Look make-Up
Retford International College

Total Look make-Up (Character)
Retford International College

MARKETING

BA (Hons) Advertising with marketing
London College of Music and Media

BA (Hons) Film Studies and Marketing (Modular)
Bolton Institute of Higher Education

BA (Hons) Media and Marketing
Farnborough College of Technology

BA (Hons) Media and Popular Culture with marketing Communications
University College Northampton

BSc in Marketing with Media Communications
University of Buckingham

HNC in Marketing
Central College of Commerce

HND in Business (Advertising and Marketing Communications)
West Herts College

HND in marketing
Central College of Commerce

Scottish Group Award in Communication and Media (Higher) Marketing
Glasgow College of Building and Printing

POST PRODUCTION

HNC in Broadcast Post-Production
Ravensbourne College of Design and Communication

PRODUCING

MA in Film and Television (Producing)
National Film and Television School

MA in Producing Film and Television
Royal Holloway

SCREENWRITING

BA (Hons) Screenwriting (Media Practice Programme)
Central Lancashire University

MA in Feature Film Screenwriting
Royal Holloway

MA in Film and Television Screenwriting
National Film and Television School

MA in Screenwriting
London College of Printing

MA in Screenwriting (Fiction)
Leeds Metropolitan University
Northern Film School

PgCert PgDip and MA in Screen writing
Liverpool John Moores University

Screenwriting
North Kensington Video-Drama Project

SCRIPTWRITING

BA (Hons) Media Performance with Scriptwriting
University of Luton

BA (Hons) Scriptwriting for Film and Television

Bournemouth Media School

MA in Television Scriptwriting
De Montfort University Leicester

PgDip and MA in Television and Radio Scriptwriting
University of Salford

PgDip in Dramatic Writing
University of Sussex

Scriptwiring
Hulme Adult learning Centre

SET DESIGN

BA (Hons) Technical Arts: Design
Wimbledon School of Arts

Setcrafts Apprenticeship Training Scheme
FT2-Film and Television Freelance Training

SOUND

Audion – Max Q
Line Out

BA (Hons) Another Subject with Sound and Music Recording
London College of Music and Media

BA (Hons) Digital Arts with Sound and Music Recording
London College of Music and Media

BSc (Hons) Audio Technology

University of Salford
Southampton Institute

BSc (Hons) Creative Music and Sound Technology
Leeds Metropolitan University

Certificates, Diplomas and Degrees in Audio and Multimedia
S and E Technology College

Foundation Sound and Vision
East Surrey College

HND in Professional Sound and Video Technology
University of Salford

MA in Film and Television (Screen Sound)
National Film and Television School

MA in Sound Design for the Moving Image
Bournemouth Media School

National Certificate in Sound Recording and Media production
North Glasgow College

National Diploma in Media (Audio and Moving Image)
North West Institute of Further and Higher Education

National Diploma in Media Production (Audio)
Lewisham College

National Diploma in Media Production (Audio and Moving Image)
Cambridge Regional College
Darlington College of Technology

Oxford College of Further Education

Sound Engineering Studies
Perth College

(2) Further Education Colleges and Universities offering a wide variety of film-related subjects – all colleges in section one are listed in alphabetical order

ABERDEEN COLLEGE, Gallowgate, Aberdeen, Scotland AB25 1BN
www.abcol.ac.uk 01224 612023

AABBINGDON COLLEGE, Northcourt Road, Abingdon, Oxford OX14 1NN 01235 555585 Email mgaston@abingdoncollege.ac.uk

ACCRINGTON AND ROSSENDALE COLLEGE, Division of Academic and Lifelong Learning, Sandy Lane, Accrington, Lancashire BB5 2AW. 01254 354219

AFECT
(Advancement of Film Education Charitable Trust), 52a Waltham Grove, London SW6 1QR
020 7609 2992

AMERSHAM AND WYCOMBE COLLEGE Stanley Hill, Amersham
Bucks HP7 9HN
www.amersham.ac.uk 01494 735555

ANGUS COLLEGE, Keptie Road, Arbroath, Angus, DD1 3EA.
www.angus.ac.uk
01241 432600

THE ARTS INSTITUTE AT BOURNEMOUTH, School of Media, Wallisdown, Poole, Dorset BH12 5HH. www.arts-inst-boutnemouth.ac.uk. 01202 363320

ARRTS INTERNATIONAL, Highfield Grange, Bubwith, North Yorks YO8 6DP. www.arrts.co.uk. 01757 288088.
AYR COLLEGE, Dam Park, Ayr, South Ayrshire KA8 OEU. 0800 199798

BARKING COLLEGE, Dagenham Road, Romford, Essex RM7 PXU
www.barking-coll.ac.uk 01708 770000

BARNET COLLEGE, Wood Street, High Barnet, Herts EN5 4AZ
www.barnet.ac.uk 020 8440 6321

BARNFIELD College, New Bedford Road, Luton, Beds LU2 7BF. 01582 569 627.

BARNSLEY COLLEGE, Honeywell Site, Honeywell Lane, Barnsley, South Yorkshire, S75 1BP. www.barnsley.ac.uk 01226 216216.

BARNSLEY COLLEGE, New Electric Theatre Studios, PO Box 266, Church Street, Barnsley, South Yorkshire, S70 2YW. www.barnsley.ac.uk 01226 216129

BARNSLEY COLLEGE, Old Mill lane Site, Church Street, Barnsley, S70 2AX. www.barnsley.ac.uk. 01226 730191.

BASINGSTOKE COLLEGE OF TECHNOLOGY, School Of Design and Communications, Worting Road, Basingstoke, Hampshire, RG21 1TN. 010256 306228.

UNIVERSITY OF BATH, Claverton Down, Bath, BA2 7AY. www.bath.ac.uk. 01225 323491.

BEDFORD COLLEGE OF HIGHER EDUCATION, Cauldwell Street, Bedford, MK42 9AH. www.bedford.ac.uk. 01234 291525

BELFAST INSTITUTE FOR FURTHER AND HIGHER EDUCATION, Ormeau Avenue, Ravenhil Road, Belfast, BT6 4GZ. 028 9026 537.

BELL COLLEGE, Almada Street, Hamilton, Lanarkshire ML3 OJB. www.bell.ac.uk. 01698 283100

BEVERLEY COLLEGE, Gallows Lane, Beverley, East Yorkshire, HU17 7DT. www.beverleycollege.ac.uk 01482 868362.

BEXLEY COLLEGE, St Josephs Campus, 269 Woolwich Road, Abbeywood, London SE2 OAR. www.bexley.ac.uk. 01322 404277.

BLACBURN COLLEGE, The Media Centre, Fielden Street, Blackburn Lancashire, BB2 1LH. 01254 292167.

BLACKPOOL AND THE FYLDE COLLEGE, Ashfield Road, Bispham, Blackpool FY2 OHB. www.blackpool.ac.uk. 01253 352352.

BLACKPOOL AND THE FYLD COLLEGE, Central Blackpool Campus, Palatine Road, Blackpool, Lancs FY1 4DW www.blackpool.ac.uk. 01253 352352.

BLAKE COLEGE, 162 New Cavendish Street, London W1M 7FJ. www.blake.ac.uk. 020 7636 0658.
BOLTON COLLEGE, Manchester Road, Bolton, BL2 1ER. www.bolton-college.ac.uk. 01204 531411.

BOLTON INSTITUTE OF HIGHER EDUCATION Chadwick Campus, Chadwick Street, Bolton BL2 1JW www.asc.bolton.ac.uk/human/film/filmhome.htm 01204 528851

BORDERS COLEGE, Thorniedean, Melrose Road, Galashiels, TD1 2AF. www.borderscollege.ac.uk. 01896 756440.

BOURNEMOUTH MEDIA SCOOL, Bournemouth University, Talbot Campus, Fern Barrow, Poole Dorset BH12 5BB. Media.Bournemouth.ac.uk 01202 595745.

BRADFORD UNIVERSITY, Department of Electronic Imaging and Media Communication, Bradford, West Yorkshire BD7 1DP www.eimc.brad.ac.uk 01274 235748

BRIDGEWATER COLLEGE, Bath Road, Bridgewater, Somerset, TA6 4PZ. 01278 455464.

BRIGHTON COLLEGE OF TECHNOLOGY, Pelham Street, Brighton, BN1 4FA. www.bricoltech.ac.uk. 01273 667788.

BRITISH ANIMATION TRAING SCHEME, Museum of the Moving Image, South Bank, Waterloo, London SE1 8XT. 020 7815 1346.

BROOKLANDS COLLEGE, Media Section, Heath Road, Weybridge, Surrey, KT13 8TT. www.brooklands.ac.uk. 01932 797726.

BRUNEL UNIVERSITY, Department of Human Sciences, Uxbridge, Middlesex UB8 3PH www.brunel.ac.uk/depts/hs 01895 274000

UNVERSITY OF BUCKINGHAM, School of Languages Hunter Street, Buckingham, MK18 1EG www.buckingham.ac.uk 01280 814080

BURNLEY COLLEGE, Ormerod Road, Burnley, Lancs BB11 2RX. 01282 436111.

CALDERDALE COLLEGE, Halifax School of Integrated Arts, Francis Street, Halifax, HX1 3UZ. 01422 399399.

CAMBRIDGE ARTS AND SCIENCES, School of Arts and Design, 13-14 Round Church Road, Cambridge, CB5 8AD. www.ccg-uk-com. 01223 314431.

CAMBRIDGE REGIONAL COLLEGE, Kings Hedges Road, Cambridge, CB4 2QT. www.camre.ac.uk. 01223 418200.

UNIVERSITY OF CAMBRIDGE, Board of Continuing Education, Madingley Hall, Madingley, Cambridge, CB3 8AQ. www.cont-ed.cam.ac.uk. 01954 280226.

CANTERBURY CHRIST CHURCH UNIVERSITY COLLEGE, Canterbury, Kent, CT1 1QU. www.cant.ac.uk. 01277 767700.

CANTERBURY COLLEGE, New Dover Road, Canterbury, Kent CT1 3AJ. www.cant-col.ac.uk. 01277 811341.

CARDONALD COLLEGE, 690 Mosspark Drive, Cardonlad, Glasgow, G52 3AY. www.cardonald.ac.uk. 0141 272 3242.

CARMARTHENSHIRE COLLEGE, Graig Campus, Sandy Road, Llanelli, SA15 4DN. www.ccta.ac.uk. 01554 748000.

CARSHALTON COLLEGE, Media, Arts and Design Centre, Nightingale Road, Carshalton, Surrey, SM5 2EJ. www.carshalton.ac.uk. 020 87706825.

CAUSEWAY INSTITUTE OF FURTHER AND HIGHER EDUCATION, 2 Colerain Road, Ballymoney, County Antrim, Northern Ireland, BT52 6BP. www.causeway.ac.uk. 028 2766 0419.

CENTRAL COLLEGE OF COMMERCE, 300 Cathedral Street, Glasgow, G1 2TA. 0141 52 3941.

CENTRAL ENGLAND IN BIRMINGHAM, UNIVERSITY OF, Perry Barr Birmingham B42 2SU. www.uce.ac.uk 0121 331 5719

CENTRAL LANCASHIRE UNIVERSITY, Faculty of Design and Technology, Department of Art and fashion, Victoria Building, Preston Lancashire PR1 2HE www.uclan.ac.uk 01772 893199

CHELMSFORD COLLEGE, Moulsham Street, Chelmsford, CM2 OJO. 01242 543236.

CHICHESTER COLLEGE OF ARTS SCIENCE AND TECHNOLOGY, Westgate Fields, Chichester, West Sussex, PO19 1SB. www.chichester.ac.uk. 01243 786321.

CIRENCESTER COLLEGE, Stroud Road, Cirencester, Glos GL7 1XA. 01285 640994.

CITY AND ISLINGTON COLLEGE, Islington Campus, 444 Camden Road, London N7 0SP. www.candi.ac.uk. 020 7700 8600.

CITY COLLEGE MANCHESTER, Arden Centre, Northenden, Manchester, M23 ODD. www.multimedia.ccm.ac.uk.

CITY COLLEGE MANCHESTER, Sale road, Northenden, Manchester, M23 ODD. www.ccm.ac.uk. 0161 957 1749.

CITY COLLEGE NORWICH, Ipswich Road, Norwich, NR2 2LJ. www.ccn.ac.uk. 01603 773269.

CITY OF BRISTOL COLLEGE, Ashley Down Road, Ashley Down, Bristol, BS7 9BU. www.cityofbristol,ac.uk. 0177 904 5143.

CITY OF WESTMINSTER COLLEGE, 25 Paddington Green, London W2 1NB. 020 7258 2789.

CLARENDON CITY COLLEGE, New College, Nottingham, The Adams Building, Stoney Street, Lace Market, Nottingham, NG1 1NG. www.cnc.ac.uk. 0115 910 4665.

CLEVELAND COLLEGE OF ART AND DESIGN, Green Lane, Linthorpe, Middlesborough, TS5 7RJ. www.ccad.ac.uk. 01642 288000.

CLYDEBANK COLLEGE, Kilbowie Road, Clydebank G81 2AA. www.clydebankcoollege.ac.uk. 0141 952 7771.

COATBRIDGE COLEGE, Kildoman Street, Coatbridge, Lanarkshire ML5 3LS. 01236 422316.

COLCHESTER INSTITUTE, Sheepen Road, Colchester, CO3 3LL. 01206 518144.

COLEG GLAN HAFREN, Trowbridge Road, Rumney, Cardiff, CF3 1XZ. www.glan-hafren-ac-uk. 029 2025 0319.

COLEG MENAI, Bangor, Gwynedd LL55 2TP. 01248 370125.

CORNWALL COLLEGE, Pool, Redruth,Cornwall, TR15 3RD. www.cornwall.ac.uk. 01209 611611.

CRAVEN COLLEGE, High Street, Skipton, North Yorkshire, BD23 1JY. 01756 791411.

CRAWLEY COLLEGE, College Road, Crawley, West Sussex, RH10 1NR. 01293 442311.

CRICKLADE COLLEGE, Charlton Road, Andover, Hampshire, SP10 1EJ. www.cricklade.ac.uk. 01264 334222.

CUMBRIA COLLEGE OF ART AND DESIGN, Brampton Road, Carlisle, Cumbria, CA3 9AY. www.cumbriacad.ac.uk. 01228 400300.

CYFLE CYF, Gronant, Penrallt Isaf, Gwynedd, LL5 1NW. www.cyfle.co.uk. 01286 671000.

CYFLE CYF, Ty Critchon, 11-12 Mount Street Square, Caerdydd CF1 6EE. www.cyfle.co.uk. 029 2046 5533..

DARLINGTON COLLEGE OF TECHNOLOGY, Cleveland Avenue, Darlington, co. Durham DL3 7BB. 01325 503229

DELEMAR ACADEMY OF MAKE-UP, 52A Waltham Grove. Fulham, London, SW6 1QR. www.themake-upcentre.co.uk 020 7381 0213.

DE MONTFORT UNIVERSITY LEICESTER , Faculty of Art and Design, The Gateway, Leicester LE1 9BH. www.dmu.ac.uk 0116 257 7507

DERBY, UNIVERSITY OF Film and Video Department, Green Lane, Derby DE1 1RX Vertigo.derby..ac..uk/hitchcock

DUDLEY COLLEGE, Mons Hill Site, 11 Wrens Hill Road, Dudley, West Midlands DY1 3SB. 01384 363192.

DUMFRIES AND GALLOWAY, Dumfries and Galloway, DG1 3QZ. www.dumgal.ac.uk 01387 243893

DUNCAN OF JORDANSTONE College of Art and Design, University of Dundee, School of Television and Imaging, perth Road, Dundee DD1 4HT www.imaging.dundee.ac.uk 01382 345250

DUNDEE COLLEGE, Media Section, 30 Constitution Road Centre, Dundee, DD3 6TB. 01382 834834.

DUNSTABLE COLLEGE, Kingsway, Dunstable, LU5 4HG. 01582 477776.

EAST ANGLIA, UNIVERSITY OF, School of English and American Studies, Norwich, NR4 7TJ. www.uea.ac.uk 01603 593428

EAST ANTRIM INSTITUTE OF FURTHER EDUCATION, 400 Shore Road, Newtownabbey, County Antrim Northern Ireland BT37 9RS. 028 9086 4331.

EAST DURHAM AND HOUGHALL COMMUNITY COLLEGE, Peterlee, Co. Durham SR8 1NU. 0191 518 2000.

EAST SURREY COLLEGE, Media Division, Gatton Point North, Claremont Road, Redhill, Surrey, RH1 2JX. www.esc.org.uk. 01737 772611.

EASTBOURNE COLLEGE OF ARTS AND TECHNOLOGY, ECAT House, Cross Levels Way, Eastbourne, BN21 2UF. 01323 644711.

EDINGBURGH COLLEGE OF ART School of Visual Communication, 74 Lauriston Place, Edinburgh, EH3 9DF. www.eca.ac.uk 0131 221 6138

EDINBURGH VIDEO TRAINING COMPANY LIMITED Unit 22, John Cotton Centre, 10 Sunnyside, Edinburgh, EH7 5RA. 0131 652 1206.

EPPING FOREST COLLEGE, Borders lane, Loughton, Essex, 1G10 3SA. 020 8508 8311.

FAREHAM COLLEGE, Bishopsfield Road, Fareham, Hampshire, PO14 1NH. www.fareham.ac.uk. 01329 815346.

FARNBOROUGH COLLEGE OF TECHNOLOGY, School of Media and Visual Arts, Boundary Road, Farnborough, Hampshire GU14 6SB. www.farn-ct.ac.uk. 01252 407 270.

FIRST FILM FOUNDATION, 9 Bourlet Close, London W1P 7PJ www.firstfilm.co.uk. 020 7580 2111.

Ft2-FILM AND TELEVISION FEELANCE TRAINING, Fourth Floor, Warwick House, 9 Warwick Street, LONDON W1R 5RA. www.ft2.org.uk 020 7734 5141.

GLAMORGAN CENTRE FOR ART AND DESIGN, Glyntaff Road, Glyntaff, Pontypridd, Mid Glamorgan CF37 4AT. 01443 663309

GLASGOW COLLEGE OF BUILDING AND PRINTING, 60 North Hanover Street, Glasgow, G1 2BP. www.gchp.ac.uk o141 332 969.

GOLDSMITHS COLLEGE, UNIVERSITY OF LONDON, Department of Media and Communications, New Cross, London SE14 6NW. www.goldsmiths.ac.uk 020 7919 7600

GRIMSBY COLLEGE, Nuns Corner, Grimsby NE Lincolnshire DN34 5BQ. www.grimsby.ac.uk. 01472 311222.

GUILDFORD COLLEGE, Stoke Park, Guildford, Surrey, GU1 1EZ. 01483 448500.

HACKNEY COMMUNITY COLLEGE, Shoreditch Campus, Falkirk Street, London N1 6HQ. www.comm-coll-hackney.ac.uk. 020 7613 9123.

HAMMERSMITH AND WEST LONDON COLLEGE, Gliddon Road, Barons Court, London W14 9BL. www.hwlc.ac.uk. 020 8741 1688.

HARLOW COLLEGE, Velizey Avenue, Harlow, Essex. CM20 3LH. www.harlow-college.ac.uk 01279 868000

HAVERING COLLEGE OF FURTHER AND HIGHER EDUCATION,
Havering Acadmey of Arts and Sciences, Ardleigh Green Road, Hornchurch, Essex RM11 2LL. www.havering-college.ac.uk 01708 455011.

HENLEY COLLEGE, COVENTRY, Department of Hospitality and Creative Studies, Henley Road, Bell Green, Coventry CV2 1ED. www.henley-cov.ac.uk. 024 7667 6300.

HEREWARD COLLEGE, Bramston Crescent, The Hil Lane, Coventry CV4 9SW. 024 7646 1231

THE HENLEY COLLEGE, Deanfield Avenue, Henley, Oxon, RG9 1UH. www.henleycol.ac.uk. 01491 579988.

HOXTON HALL, 130 Hoxton Street, London N1 6SH. www.hoxtonhall.dobsol.co.uk 020 7684 0060.

HUDDERSFIELD, UNIVERSITY OF, School of Music and Humanities, Queensgate, Huddersfield, West Yorkshire HD1 3DH www.hud.ac.uk 01484 472359

HULL COLLEGE, Riley Centre, Parkefield Drive, Hull, HU3 6TE. www.hull-college.ac.uk 01482 351228.

HULME ADULT LEARNING CENTRE, Hulme Centre, Stretford Road, Manchester M15 5FQ. 0161 226 8411.

HULME ADULT LEARNING CENTRE, Hulme Walk, Hulme, Manchester M15 5FQ. . 0161 226 8411.

INTERMEDIA FILM AND VIDEO (NOTTINGHAM) Ltd, 19 Heathcote Street, Nottingham NG1 3AF, www.intermedianotts.co.uk. 0115 955 6909.

INVERNESS COLLEGE, 3 Longman Road, Longman South, Inverness 1V1 1SA. 01463 273353.

JAMES WATT COLLEGE, Finnart Street, Greenock, Renfrewshire, PA16 8HF. www.jameswatt.ac.uk 01475 724433.

JAMES WATT COLLEGE, North Ayrshire Campus, Lauchlan way, Kilwinning, North Ayrshire www.jameswatt.ac.uk 01294 555314.

JEWEL AND ESK VALLEY COLLEGE, Milton Road Centre, 24 Milton Road East, Edingburgh EH15 29. www.jevc.ac.uk 0131 657 7308

KENT INSTITUTE OF ART AND DESIGN, Maidstone College, Oakwood Park, Maidstone, Kent ME16 8AG. 01622 757286

KING ALFREDS WINCHESTER, Sparkford Road, Winchester SO22 4NR 01962 827368

KINGSTON COLLEGE, Division of Media and Performing Arts, Department of Design Studies, Kingston Hall Road, Kingston Upon Thames, Surrey KT1 2AQ. 020 8 268 2897.

LAMBETH COLLEGE, Vauxhall centre, Belmore Street, Wandsworth Road, London SW8 2JY. www.lambethcollege.ac.uk 020 7501 5618.

LEEDS COLLEGE OF ARTS AND DESIGN, Jacob Kramer Buildings, Blenheim walk, Leeds, LS2 9AQ. www.leeds-art.ac.uk 0113 202 8136.

LEEDS COLLEGE OF MUSIC, 3 Quarry Hill, Leeds, LS2 7PD. www.lcm.ac.uk 0113 222 3400.

LEEDS METROLPOLITAN UNIVERSITY, Calverley Street, Leeds, LS1 3HE. www.lmu.ac.uk 0113 283 3113.

LEICESTER COLLEGE, St Margaret's Campus, Grafton Place, St John Street, Leicester, LE1 3WL. www.leicestercollege.ac.uk 0116 224 2002.

LEWISHAM COLLEGE, Tressilian Building, Lewisham Way, London SE4 1UT. www.lewisham.ac-uk/college 020 8692 0353.

LIGHT HOUSE, The Chubb Building, Fryer Street, Wolverhampton, WV1 1HT. www.light-house.co.uk 01902 716055

LIGHTHOUSE, 9-12 Middle Street, Brighton, BN1 1AL. www.lighthouse.org.uk 01273 384258.

LINE OUT, Fosse Arts Centre, Mantle Road, Leicester, LE3 5HG. 0116 262 1265.

LIVERPOOL COMMUNITY COLLEGE, Professional media, The Arts Centre, 9 Myrtle Street, Liverpool, L7 7JA. www.liv-coll.ac.uk 0151 252 4328

LIVERPOOL JOHN MOORES UNIVERSITY, School of Media, Critical and Creative Arts, Dean Walters Building, St James Road, Liverpool L1 7BR www.livjm.ac.uk 0151 231 5052

LONDON COLLEGE OF MUSIC AND MEDIA, Thames Valley University, St Mary's Road, Ealing, London W5 5RF elgar.tvu.ac.uk 020 8231 2022

LONDON FILM MAKING ACADEMY, The Old Church, 52a Waltham Grove, London SW6 1QR. www.londonfilmacademy.com 020 7386 7711.

LOUGHBOROUGH COLLEGE, Community and General Education, Radmore road, Loughborough, Leicestershire, LE11 3BT. www.loucoll.ed.uk 01509 215831.

LUTON, UNIVERSITY OF, Faculty of Humanities, Department of Media Arts, 75 Castle street, Luton LU1 3AJ www.luton.ac.uk 01582 489162

MANCHESTER, UNIVERSITY OF, Department of Drama, Oxford Road, Manchester, M13 9PL www.art.man.ac.uk/drama/ 0161 275 3964

MERTON COLLEGE, Central Road, Mordern, Surrey SM4 5SE. www.merton.ac.uk 020 8408 8663.

MORLEY COLLEGE, 61 Westminster Bridge Road, London SE1 7HT. www.morleycollege.ac.uk 020 7450 9235.

MOTHERWELL COLLEGE, Dalzell Drive, Motherwell, ML1 2DD. www.motherwell.ac.uk. 01698 232652.

NAPIER UNIVERSITY, Department of Print media, Publishing and Communication, Craighouse Road, Edinburgh EH10 5LG www.napier.ac.uk 0131 455 6165

NATIONAL FILM AND TELEVISION SCHOOL, Beaconsfield Studios, Station Road, Beaconsfiled, Bucks, HP9 1LG www.nftsfilm-tv.ac.uk 01494 671234

NERVE CENTRE, 7-8 Magazine Street, Derry City, Northern Ireland, BT48 6HJ. www.nerve-centre.org.uk 028 7126 0562.
NESCOT (NORTH EAST SURREY COLLEGE OF TECHNOLOGY), Reigate Road, Ewell, Surrey, KT17 3DS. www.nescot.ac.uk. 020 8394 3192.

NEW COLLEGE DURHAM, Nevilles cross Centre, Darlington Road, Durham, DH1 4SY www.newdur.ac.uk 0191 375 4118

NEW COLLEGE SWINDON, Helston Road, Park North, Swindon, SN3 2LA. 01793 611470.

NEW PRODUCERS ALLIANCE, 9 Bourlet Close, London W1W 7BP. www.newproducer.co.uk 020 7580 2480.

NEWBURY COLLEGE OF FURTHER EDUCATION, Oxford Road, Newbury Berks RG13 1PQ. 01635 37000.

NEWCASTLE COLLEGE, School of Arts and Design, John marley Centre, Wickham View, Newcastle Upon Tyne NE15 6T. www.ncl-coll.ac.uk 0191 200 4725.

NEWCASTLE COLLEGE, School of Music and Performing Arts, Rye Hill Campus, Scotswood Road, Newcastle upon Tyne, NE4 7SA. www.ncl-coll.ac.uk 0191 200 4684.

NEWCASTLE UNDER LYME COLLEGE, Liverpool Road, Newcastle Under Lyme, Staffordshire, ST5 2DF. 10782 254325.

NORTHBOROOK COLLEGE, Litlehampton Road, Goring by Sea, Worthing, West Sussex BN12 6NU www.northbrok.ac.uk 01903 606001

NORTH EAST WALES INSTITUTE OF HIGHER EDUCATION, North wales School of Art and Design, 49 Regent Street, Wrexham, LL11 1PF. www.newi.ac.uk/nwsad 01978 293552.
NORTH EAST WORCESTERSHIRE COLLEGE, Redditch Campus, Peakman Street, Redditch, Worcs B98 8DW. 01527 570020

NORTH WEST INSTITUTE OF FURTHER AND HIGHER EDUCATION, Strand Road, Londonderry BT48 7BY www.nwifhe.ac.uk 028 7126 6711

NORTH GLASGOW COLLEGE, 110 Flemington Street, Glasgow, G21 4BX. www.north-gla.ac.uk. 0141 558 9001.

NORTH HERTFORDSHIRE COLLEGE, Stevenage Centre, Monkswood way, Stevenage, Herts, SG1 1LA. 01462 424242.

NORTH KENSINGTON VIDEO DRAM PROJECT, 1 Thorpe Close, London W10 5XL. 020 8964 2641.

NORTH NOTTINGHAMSHIRE COLLEGE, Carlton Road, Worksop, Notts S81 7HP. 01909 473561.

NORTH OXFORDSHIRE COLLEGE, Broughton Road, Banbury, Oxon, OX16 9QA. www.northox.ac.uk 01295 252221.

NORTHERN FILM SCHOOL Leeds Metropolitan University, 2 Queens Square, Leeds LS2 8AF. www. lmu.ac.uk/hen/aad/nfs 0113 283 1900

NORTH TYNESIDE COLLEGE, Embleton Avenue Wallsend Tyne and wear NE28 9NJ www.ntyneside.ac.uk 0191 229 5000

NORTH WARWICKSHIRE AND HINCKLEY COLLEGE, New Media and Performing Arts Centre, London Road, Hinckley, Leicestershire LE10 1HQ. www.nwarks-hinckley.ac.uk. 024 7634 9321.

NORWICH SCHOOL OF GRAPHIC ART AND DESIGN, St George Street, Norwich NR3 1BB www.nsad.ac.uk 01603 610561

NOTTINGHAM TRENT UNIVERSITY, Burton Street, Nottingham NG1 4BU www.ntu.ac.uk 0115 848 4850

OAKLANDS COLLEGE, Art, Media and Performing Arts, St Albans City Campus, St Peters Road, St Albans, Herts, AL1 3RX. 01707 737000.

THE OLDHAM COLLEGE, School of Visual and performing Arts, Rochdale Road, Oldham, OL9 6AA. www.oldham.ac.uk. 0161 785 4052.

OXFORD COLEGE OF FURTHER EDUCATION, Faculty of Arts, Oxpens Road, Oxford, OX1 1SA. www.oxfe.ac.uk/ocfe 01865 269482.

PEMBROKESHIRE COLLEGE, Merlins Bridge, Haverfordwest, Pembrokeshire, SA61 1SZ. www.pembrokeshire.ac.uk 01437 765247.

THE PEOPLE COLLEGE NOTTINGHAM, Maid Marian Way, Nottingham, NG1 6AB. www.peoples.ac.uk 0115 912 3500.

PERTH COLLEGE, Faculty of Arts, Goodyburn East, Crieff Road, Perth, Perthshire PH1 2NX. www.uhi.ac.uk/perth 01738 621171.

PLYMOUTH COLLEGE OF ART AND DESIGN, Tavistock Place, Plymouth, PL4 8AT. www.pcad.plym.ac.uk 01752 203434.

PONTYPRIDD COLLEGE, YNYS Terrace, Rhydyfelin, Pontypridd, Mid Glamorgan CF37 5RN. www.pontypridd.ac.uk. 01443 663090.

QUEEN ELIZABETH'S FOUNDATION TRAINING COLLEGE, Leatherhead Court, Leatherhead, KT22 OBN. 01372 841100.

RAVENSBOURNE COLEGE OF DESIGN AND COMMUNICATION, Walden Road, Chislehurst, Kent BR7 5SN www.ravc.ac.uk 020 8289 4900

REDBRIDGE COLLEGE, Little Heath, Romford, Essex, RM6 4XT. 020 8548 7400.

RETFORD INTERNATIONAL COLLEGE, 11 Grove Street, Retford, Notts, DN22 6PJ. (Make up and design) 01777707371.

RICHMOND UPON THAMES COLLEGE, Media Education Workshop, Egerton Road, Twickenham, TW2 7SJ. 020 8607 8217.

ROYAL COLLEGE OF TECHNOLOGY, Kensington Gore, London SW7 2EU www.rca.ac.uk 020 7590 4512

ROYAL HOLLWAY, University of London, Department of Media Arts Egham Surrey TW20 OEX www.media.dr.rhul.ac.uk 01784 443734

SALISBURY COLLEGE, Southampton Road, Salisbury, Wilts SP1 2LW. www.salcol.com 01722 344344

ST AUSTELL COLLEGE, Trevarthian Road, St Austell, Cornwall, PL25 4BU. 01726 67911.

ST HELENS COLLEGE, School of Arts, Media and Design, Town Centre Campus, Brook Street, St Helens, Merseyside, WA10 1PZ. www.sthelens.ac.uk. 01744 623253.

ST VINCENT COLLEGE, Mill Lane, Gosport, Hants PO12 4QA. 023 9258 8311.

UNIVERSITY OF SALFORD, Salford M5 4WT. www.salford.ac.uk 0161 295 5321.

SANDWELL COLLEGE, Lakeside Studios, Smethwick Campus, Crocketts Lane, Smethwick, West Midlands B66 3BU. www.sandwell.ac.uk 0121 253 6341.

SELBY COLLEGE, Department of Art, Design and Media, Abbots Road, Selby YO8 8AT. www.selbycollege.co.uk. 01757 211000.

THE SHEFFIELD COLLEGE, Norton Centre, Dyche Lane, Sheffiled S8 8BR. www.sheffcol.ac.uk 0114 260 2300.

SIF (Sheffield Independent Film), 5 Brown Street, Sheffield, S1 2BS. 0114 272 0304.

SOMERSET COLLEGE OF ARTS AND TECHNOLOGY, Wellington Road, Taunton, Somerset TA1 5AX. www.somerset.ac.uk 01823 366311.

SOUTHAMPTON INSTITUTE, East Park Terrace, Southampton, SO14 OYN www.solent.ac.uk

SOUTH CHESHIRE COLLEGE, Creative and Performing Arts, Dane Bank Avenue, Crewe, Cheshire, CW2 8AB. www.s-cheshire.ac.uk 01270 654654.

SOUTH DEVON COLLEGE, Newton Road, Torre, Torquay, Devon,TQ2 5BY. 01803 406253

SOUTH DOWNS COLLEGE OF FURTHER EDUCATION, College Road, Havant, Hampshire, PO7 8AA. 023 9279 7979.

SOUTH EAST ESSEX COLLEGE, Carnarvon Road, Southend on Sea, Essex SS2 6LS. www.se-essex-college.ac.uk. 01702 220400.

SOUTH KENT COLLEGE, The Media Centre, Maison Dieu Road, Dover, Kent, CT16 1DH. 01304 244355.

SOUTH THAMES COLLEGE, Faculty of Creative and Technological Studies, Wandsworth High Street, London SW18 2PP. www.south-thames.ac.uk 020 8918 7315.

SOUTH TRAFFORD COLLEGE, Manchester Road, Timperley, Cheshire, WA14 5PQ. 0161 952 4733.

SOUTHGATE COLLEGE, High Street, London N14 6BS. www.southgate.ac.uk 020 8982 5064.

SOUTHWARK COLLEGE, Surrey Docks Centre, Drummond Road, London SE16 4EE. www.southwark.ac.uk. 020 7815 1500.

STAFFORDSHIRE, UNIVERSITY OF, School of Art and Design, College Road, Stoke on Trent ST4 2DE www.staffs.ac.uk 01782 294565

STEVENSON COLLEGE, Bankhead Avenue, Sighthill, Edinburgh, EH1 4DE. www.stevenson.ac.uk 0131 535 4600.

STOCKPORT COLLEGE OF FURTHER AND HIGHER EDUCATION, Wellington Road South, Stockport, Cheshire SK1 3UQ. www.stockport.ac.uk 0161 958 3168.

STRATFORD UPON AVON COLLEGE, The Willows North, Stratford Upon Avon Warwickshire CV37 9QR. 01789 266254.

STROUD COLLEGE OF FURTHER EDUCATION, Stratford Road, Stroud, Glos, GL5 4AH. 01453 763424.

SUNDERLAND, UNIVERSITY OF, School of Arts, Design and Media, Ashbourne House, Backhouse Park, Ryhope Road, Sunderland SR2 7EF www.sunderland.ac.uk 0191 515 2112

SURREY, ROEHAMPTON, UNIVERSITY OF, School of Arts, Digby Stuart Colege, Roehampton Lane, London SW15 5PU www.roehampton.ac.uk 020 8392 3230

SURREY INSTITUE OF ART AND DESIGN, Falkner Road, The Hart, Farnham, Surrey GU9 7DS www.surrart.ac.uk 01252 722441

SUTTON COLDFIELD COLEGE OF FURTHER EDUCATION, Lichfield Road, Sutton Coldfield, Birmingham, B74 2NW. www.sutcol.ac.uk 0121 355 5671.

SWANSEA COLLEGE, Tychoch Road, Swansea, SA2 9EB. www.swancoll.ac.uk. 01792 284023.

SWANSEA INSTITUTE OF HIGHER EDUCATION, Faculty of Humanities, Education and Healthcare, Townhill Road, Swansea, SA2 OUT www.sihe.ac.uk 01792 481000

SWINDON COLLEGE, School of Art and Design, Euclid Street, Swindon, Wiltshire, SN21 2JQ. www.swindon-college.ac.uk. 01793 498490.

TAMESIDE COLLEGE, Clarendon Road, Hyde, Cheshire, SK14 2JZ. www.tamesidecollege.ac.uk 0161 908 6800.

TEESIDE TERTIARY COLLEGE, Faculty of Creative Arts and Media, Longlands Campus, Douglas Street, Middlesborough, Cleveland TS4 2JW. www.ttc.ac.uk 01642 275000.

TELFORD COLLEGE OF ARTS AND TECHNOLOGY, Haybridge Road, Wellington, Telford, Shropshire TF1 2NP. www.tcat.ac.uk 01952 642211.

THOMAS DANBY COLLEGE, Rounday Road, Leeds LS7 3BG. www.thomasdanby.ac.uk. 0113 284 6220.

TOTTON COLLEGE, Water Lane, Totton, Southampton SO40 3ZX. www.totton.ac.uk. 023 8087 4874.

TOWER HAMLETS COLLEGE, Poplar Centre, Poplar High Street, London E14 OAF. www.tower.ac.uk 020 7538 5888.

TRESHAM INSTITUTE OF FURTHER AND HIGHER EDUCATION, Windmill Avenue, Kettering, Northamptonshire. 01536 410252.

TRURO COLLEGE, College Road, Truro, Cornwall TR1 3XX. 01872 264251.

UNIVERSITY COLLEGE NORTHAMPTON, Park Campus, Broughton Green Road, Northampton, NN2 7AL www.northampton.ac.uk 01604 735500

UNIVERSITY OF GLAMORGAN, School of Humanities and Social Sciences, Pontypridd, Mid Glamorgan CF37 1DL www.glam.ac.uk 01433 480480

UNIVERSITY OF KINGSTON, School of Music, Kingston Hill, Kingston-Upon-Thames, Surrey KT2 7LB. www.kingston.ac.uk 020 8547 7149

UNIVERSITY OF LINCOLNSHIRE AND HUMBERSIDE, Hull School of Art and Design, Queens Gardens, Hull HU1 3DQ www.ulh.ac.uk 01482 440550

UNIVERSITY OF SALFORD, Salford M5 4WT www.salford.ac.uk 0161 295 5321

UNIVERSITY OF SUSSEX, Media Studies Faculty, Essex House, Brighton BN1 9RQ www.sussex.ac.uk 01273 678019

UNIVERSITY OF TEESIDE, School of Law, Arts and Humanities, Middlesborough, Cleveland, TS1 3BA www.tees.ac.uk 01642 384019

UNIVERSITY OF WALES NEWPORT, Caerleon Campus PO Box 179 Newport NP18 1YG www.newport.ac.uk 01633 430088

UNIVERSITY OF WOLVERHAMPTON, Wolverhampton Science Park, Wolverhampton WV10 9RY www.wlv.ac.uk 01902 824003

UXBRIDGE COLLEGE, Park Road, Uxbridge, Middlesex, UB8 1NQ. www.uxbridge.ac.uk 01895 853510.

VARNDEAN COLLEGE, Surrenden Road, Brighton, East Sussex, BN1 6WQ. www.varndean.ac.uk 01273 508011.

WAKEFIELD COLLEGE, Thornes Park Centre, Thornes Park, Wakefield, WF2 2QZ. www.wakcoll.ac.uk 01924 789808.

WALES, ABERYSTWYTH UNIVERSITY OF Department of Film Theatre and Television Studies, Parry Williams Building Penglais Campus Aberystwyth SY23 2AJ www.aber.ac.uk.tfts 01970 622369

WALSALL COLEGE OF ARTS AND TECHNOLOGY, Shelley Centre, Scarborough Road, Pleck, Walsall WS2 2TY. 01922 720889.

WARWICKSHIRE COLLEGE, Warwick New Road, Leamington Spa, Warwickshire, CV32 5JE. www.warkscol.ac.uk 01926 318000.

WARWICK, UNIVERSITY OF, Department of Film and Television Studies, Faculty of Arts, Coventry CV4 7AL www.warwick.ac.uk 024 7652 3511

WARRINGTON COLLEGIATE INSTITUTE, Faculty of Higher Education, Padgate Campus, Fearnhead Lane Fearnhead, Warrington WA2 ODB. www.warr.ac.uk 01925 494205

WEST HERTS COLLEGE, School of Business and Management, Hempstead Road, Watford WD17 3EZ. www.westherts.ac.uk 01923 812592.

WEST NOTTINGHAMSHIRE COLLEGE, Derby Road, Mansfield, Notts, NG18 5BH. www.westnotts.ac.uk 01623 627191.

WEST KENT COLLEGE, Brook Street, Tonbridge, Kent, TN9 2PW 01732 358101

WEST THAMES COLLEGE, London Road, Isleworth, Middlesex TW7 4HS. 020 8568 2331.

WESTMINSTER ADULT EDUCATION SERVICE, Ebury Bridge Centre, 4 Sutherland Street, London SW1 4LH. www.waes.ac.uk 020 7641 7852.

WESTMINSTER KINGSWAY COLLEGE, Battersea Park Road, London SW11 4JR. www.wesking.ac.uk 020 7556 8000.

WESTMINSTER, UNIVERSITY OF, School of Communications and Creative Studies, Harrow Campus, Watford Road, Northwick Park, Harrow HA1 3TP www.wmin.ac.uk 020 7911 5903

WIGAN AND LEIGH COLLEGE, The Media Centre, leigh Campus, Railway Road, Leigh WN7 4AH. www.wigan-leigh.ac.uk 01942 761729.

WILTSHIRE COLLEGE CHIPPENHAM, Cocklebury Road, Chippenham, Wiltshire, SN15 3QD. www.wiltscoll.ac.uk. 01249 464444.

WIMBLEDON SCHOOL OF ART, Merton Hall Road, London SW19 3QA www.wwwebguides.com 020 8408 5080

WOLVERHAMPTON COLLEGE, Wulfrun Campus, Paget Road, Wolverhampton
WV6 ODU. www.wolverhamptoncollege.ac.uk 01902 317700.

WYKE COLLEGE, Grammar School Road, Hull HU5 4NX. www.hull.ac.uk/hull/ctls-web/wyke.html 01978 311794.

YMTC (YORKSHIRE MEDIA TRAINING CONSORTIUM) 40 Hanover Square, Leeds, LS3 1BQ. (New Entrants) www.ymtc.co.uk 0113 294 4410.

YORK COLLEGE, Tadcaster Road, York YO24 1UA. www.yorkcollege.ac.uk. 01904 770294.

SHORT COURSES

There are over 170 centres in England Scotland, Wales and Northern Ireland offering short courses. These courses range from one-day events through to part time evening classes and to courses lasting for several weeks. The centres include colleges of further education and higher education, film and video workshops and commercial training centres.

For course titles, duration, entry requirements and cost, see media and multimedia short courses, which is produced jointly by the British Film Institute and Skillset. This is available from Plymbridge Distributors Ltd, Estover Road, Plymouth, Devon PL6 7PZ. Tel: 01752 202301. Priced currently at £4.50.

The short courses information may also be found on the British Film Institute website www.bfi.org.uk/mediacourses.

ABERDEEN COLLEGE, Gallowgate, Aberdeen, Scotland AB25 1BN. www.abcol.ac.uk 01224 612325.

ACADEMY OF RADIO,FILM AND TELEVISION, American Building, 79A Tottenham Court Road, London W1T 4TD. www.londonacademy.co.uk 020 84048298.

AMERSHAM AND WYCOMBE COLLEGE, Stanley Hil, Amersham, Bucks HP7 9HN. www.amersham.ac.uk 01494 735555.

THE ARIEL TRUST LTD, Floor 3, Graphic House, Duke Street, Liverpool, L1 4JR.www.arieltrust.com 0151708 0595.
ARISTA, 11 Wells Mews, London W1P 3FL. 020 7323 1775.

THE ARVON FOUNDATION, Lumb Bank, Heptonstall, Hebden Bridge, West Yorkshire HX7 6DF. www.arvonfoundation.org 01422 843714.

AVRIL ROWLANDS TV TRAINING FOR PRODUCTION ASSISTANTS, Laxford House, Poplar piece, Inkberrow, Worcs WR7 4JD. www.avriltrain.com 01368 792051.

BBC TRAINING AND DEVELOPMENT, BBC New Media Training, 35 Marylebone High Street, London W1U 4PX. www.bbctraining.co.uk 020 7208 9426.

BBC TRAINING AND DEVELOPMENT, Wood Norton, Evesham, Worcs WR11 4YB. www.bbctraining.co.uk 0870 122 0216.

BEXLEY COLLEGE, St Josephs Campus, 269 Woolwich Road, Abbeywood, London SE2 OAR. www.bexley.ac.uk 01322 404277.

BIRKBECK COLLEGE, Faculty of Continuing Education (Media Studies), 26 Russell Street, London WC1B 5DQ. www.bbk.ac.uk 020 7631 6667.

BKSTS – THE MOVING IMAGE SOCIETY, 5 walpole Court, Ealing Studios, Ealing Green, London W5 5ED. www.bksts.com 020 8584 5220

BLACKBURN COLLEGE, The Media Centre, Fielden Street, Blackburn, Lancashire BB2 1LH. 01254 682700.

BLAKE COLLEGE, 167 New Cavendish Street, London W1M 7FJ. www.blake.ac.uk 020 7636 0658.

BLAZE THE TRAIL LTD, 2nd Floor, 241 High Street, Walthamstow, London E17 7BH. www.blaze-the-trail.com 020 8520 4569.

BRADFORD COMMUNITY BROADCASTING, 2 Forster Square Bradford BD1 1DQ. www.bcb.yorks.com 01274 771677.

BRIGHTON COLLEGE OF TECHNOLOGY, Pelham Street, Brighton, BN1 4FA. www.bricoltech.ac.uk 01273 66738.

BRITISH UNIVERSITIES FILM AND VIDEO COUNCIL, 77 Wells Street, London W1T 3QJ. www.bvfc.ac.uk 020 7393 1504.

CAMBERWELL COLLEGE OF ARTS, Developments at London Institute (DALI), Peckham Road, London SE5 8UF. www.camb.linst.ac.uk. 020 7514 6311.

INIVERSITY OF CAMBRIDGE, Board of Continuing Education, Madingley Hall, Madingley, Cambridge CB3 8AQ. www.cont-ed.cam.ac.uk 01954 280226.

CENTRAL SAINT MARTINS COLLEGE OF ART AND DESIGN, Short Course Office, Southampton Row,London WC1B 4AP. www.csm.linst.ac.uk/shortcourse 020 7514 7015.

CHELSEA COLLEGE OF ART AND DESIGN, Developments at London Institute (DALI), Hugon Road, London SW6 3ES. www.chelsea.linst.ac.uk 020 7514 6311.

THE CHILDRENS FILM UNIT, South Way, Leavesden, Herts WD2 7LZ. www.btinternet.com/-cfu 01923 354656.

THE CINEMA AND FILM WORKSHOP DEPT, Chapter Arts Centre, Market Road, Canton,Cardiff CF5 1QE. 029 2031 1050.

CLINTEL INTERNATIONAL, Watton Road, Ware, Herts SG12 OAE. www.cintel.co.uk 01920 463939.

CIRENCESTER COLLEGE, Stroud Road, Cirencester, Glos GL7 1XA. 01285 640994.

CITY AND ISLINGTON COLLEGE, Islington Campus, 444 Camden Road, London N7 OSP. www.candi.ac.uk 020 7700 8600.

CITY COLLEGE BRIGHTON AND HOVE, Pelham Street, Brighton, BN1 4FA. www.cch.ac.uk. 01273 667738.

CITY EYE LTD, 1st Floor, Northam Centre, Kent Street, Northam, Southampton SO14 5SP. www.city-eye.co.uk. 023 8063 4177.

THE CITY LITERARY INSTITUTE, Stukeley Street, London WC2B 5LJ. 020 7430 0542.

CITY OF BRISTOL COLLEGE, Ashley Down Road, Ashley Down, Bristol BS7 9BU. www.cityofbristol.ac.uk. 0117 904 5143.

CITY UNIVERSITY, Courses for Adults, Northampton Square, London EC1V OHB. www.city.ac.uk/contel.cfa.htm 020 7040 8268.

CLOCKTOWER MEDIA TRAINING, Croydon Clocktower, Katherine Street, Croydon CR9 1ET. www.croydon.gov.uk/media 020 8760 5400.

COMMUNITY MEDIA ASSOCIATION, 15 Paternoster Place, Sheffiled S1 2BX. www.commedia.org.uk 0114 279 5219.

CONNECTIONS COMMUNICATIONS CENTRE, Palingwick House, 241 King Street, Hammersmith, London W6 9LP. www.cccmedia.co.uk 020 8741 1766.

CORNWALL MEDIA RESOURCE, Royal Circus Buildings, Back Lane west, Redruth, Cornwall TR15 2BT. www.mediasource.freeserve.co.uk 01209 218288.

CRICKLADE COLLEGE, Charlton Road, Andover, Hampshire SP10 1EJ. www.cricklade.ac.uk 01264 334222.

CSV DUNDEE MEDIA ACTION, Unit H, Market Mews, Broughty Ferry Road, Dundee DD1 3NH. 01382 451840.

CSV MEDIA MIDLANDS, Hollymoor Multimedia centre, Manor Park Grove, Off Tessall Lane, Northfield, Birmingham B31 5ER. 0121 683 1800.

DELAMAR ACADEMY OF MAKE-UP, 52A Waltham Grove, Fulham, London SW6 1QR. www.themake-upcentre.co.uk 020 7381 0213.

DERRY MEDIA ACCESS AND LOCAL TV NETWORK, Foyles Arts Centre, Lawrence Hill, Derry, Londonderry BT48 7NJ. 028 7137 0091.

EALING THE COLLEGE, Ealing Centre, The Green, Ealing, London W5 5EW. www.etc.ac.uk 020 8231 6006.

EAST SURREY COLLEGE, Media Division, Gaton Point North, Claremont Road, Redhill, Surrey RH1 2JX. www.esc.org.uk 01737 772611.

UNIVERSITY OF EDINBURGH, Centre for Continuing Education, 11 Buccleuch Place, Edinburgh EH8 9LW. www.cce.ed.ac.uk 0131 650 3073.

EUROSCRIPT, Screenwriters Centre Suffolk house, 1-8 Whitfield Place, London W1T 5JU www.euroscript.co.uk 020 7387 5880.

Fareham College, Bishopsfield Road, Fareham, Hampshire, PO14 1NH. www.fareham.ac.uk. 01329 815200.

FILM NAND VIDEO ACCESS CENTRE, 25A SW Thistle street lane, Edinburgh EH1 22EW. www.fva-online.com 0131 220 0220.

FOUR CORNERS FILM WORKSHOP, 113 Roman Road, Bethnal Green, London E2 OQN. www.fourcornersfilm.org.uk 020 8981 4243.

GATESHEAD COLLEGE, Durham Road, Gateshead, Tyne and wear NE9 5BN. 0191 490 2474.

GLASGOW MEDIA ACCESS CENTRE, 3rd Floor, 34 Albion Street, Glasgow G1 1LH. www.g-mac.co.uk 0141 553 2620.

HACKNEY COMMUNITY COLLEGE, Shoreditch Campus, Falkirk Street, London N1 6HQ. www.comm-coll-hackney.ac.uk 020 7613 9123.

HAMMERSMITH AND WEST LONDON COLLEGE, Gliddon Road, Barons Court, London W14 9BL. www.hwlc.ac.uk. 020 8741 1688.

HARLOW COLLEGE, Velizy Avenue, Harlow, Essex, CM20 3LH. www.harlow-college.ac.uk 01279 868000.

HOXTON HALL, 130 Hoxton Street, London N1 6SH. www.hoxtonhall.dabsol.co.uk 020 7684 0060.

HUDDERSFIELD TECHNICAL COLLEGE, New North Road, Huddersfield, HD1 5NN. 01484 348100.

HULL TIME BASED ARTS, 42-43 High Street, Hull HU1 1PS. www.timebase.org 01482 216446.

HULME ADULT LEARNING CENTRE, Hulme Walk, Hulme, Manchester M15 5FQ. 0161226 8441.

INTERMEDIA FILM AND VIDEO (NOTTINGHAM) LTD, 19 Heathcote Street, Nottingham, NG1 3AF. www.intermedianotts.co.uk 0115 95 6909.

INTERNATIONAL FILM SCHOOL WALES, University of Wales College, Newport, Caerleon Campus, PO Box 179, Newport NP18 3YG. www.ifsw.newport.ac.uk 01633 432679.

INVERNESS COLLEGE, 3 Longman Road, Longman South, Inverness, IV1 1SA. 01463 273353.

KENSINGTON AND CHELSEA COLLEGE, Hortensia Centre, Hortensia Road, London SW10 OQS. 020 7573 5346.

KILMARNOCK COLLEGE OF FURTHER EDUCATION, Holehouse Road, Kilmarnock, Ayrshire KA3 7AT. www.kilmarnock.ac.uk 01563 523501.

LAMBETH COLLEGE, Vauxhall Centre, Belmore Street, Wandsworth Road, London SW8 2JY. www.lambethcollege.ac.uk 020 7501 5259.

LEEDS ANIMATION WORKSHOP, 45 Bayswater Row, Leeds, West Yorkshire, LS8 5LF. 0113 248 4997.

LEEDS UNIVERSITY, School of Continuing Education, Leeds LS2 9JT. www.leeds.ac.uk/aed/cehome/sce.htm 0113 23 3205.

LIGHTHOUSE, 9-12 Middle Street, Brighton, BN1 1AL. www.lighthouse.org.uk 01273 384258.

LIGHTHOUSE, The Chubb Building, Fryer Street, Wolverhampton WV1 1HT. www.light-house.co.uk 01902 71605.

LONDON FILM ACADEMY, The Old Church, 52a Waltham Grove, London SW6 1QR. www.londonfilmacademy.com 020 7386 7711.

MEDIA PRODUCTION TRAINING FACILITIES, 1st Floor, 383 Brixton Road, London SW9 8BE. www.media-production.demon.co.uk 020 7737 7152.
METRO NEW MEDIA, 35 Kingsland Road, Shoreditch, London E2 8AA. www.metronewmedia.com 020 7729 9992.

MORLEY COLLEGE, 61 Westminster Bridge Road, London SE1 7HT. www.morleycollege.ac.uk. 020 7450 9235.

MOVING VISION, Wyvern Road, Newbridge on Wye, Powys, LD1 6LH. www.movingvision.co.uk. 01597 860575.

NATIONAL SHORT COURSE TRAINING PROGRAMME, National Film and Television School, Beaconsfield Studios, Station Road, Beaconsfield, Bucks HP9 1LG. www.nsftsfilm-tv.ac.uk 01494 677903.

NETWORK, 3 Harriers View, Banbury, Oxon, OX16 9JP. 01295 263273.

NEWCASTLE COLLEGE, Rye Hill Campus, Scotswood Road, Newcastle Upon Tyne NE4 7SA. www.ncl-coll.ac.uk 0191 200 4000.

UNIVERSITY OF NEWCASTLE UPON TYNE, Centre for Lifelong Learning, King George V1 Building, Queen Victoria Road, Newcastle Upon Tyne NE1 7RU. www.ncl.ac.uk 0191 222 6542.

NORTH BIRMINGHAM COLLEGE, Aldridge Road, Great Barr, Birmingham, B44 8NE. www.northbham.ac.uk 0121 325 2245.

NORTH EAST WORCESTERSHIRE COLLEGE, Redditch Campus, Peakman Street, Redditch, Worcs B98 8DW. 01527 570020.

NORTH KENSINGTON VIDEO DRAMA PROJECT, 1 Thorpe Close, London W10 5XL. 020 8964 2641.

NORTHERN FILM SCHOOL, Leeds Metropolitan University, 2 Queen Square, Leeds, LS2 8AF. www.lmu.ac.uk/hen/aad/nfs 0113 283 1900.

NORTHERN VISION MEDIA CENTRE, 4 Lower Donegall Street, Belfast BT1 2FN. www.northernvisions.org 028 9024 5495.

OXFORD FILM AND VIDEO MAKERS, The Stables, North Place, Headington, Oxford OX3 9HY. www.ofvm.org 01865 760074.

PAVILION, 2 Woodhouse Street, Leeds, LS3 1AD. www.pavilion.org.uk 0113 243 1749.

PERFORMING ARTS LABS, The Gym, Mary ward House, 5 Tavistock Place, London WC1H 9SN. 020 7387 96.

PERSONAL PRESENTATION LIMITED, Canalot Production Studios, 222 Kensal Road, London W10 5BN. www.personalpresentation.com 020 8968 0421.

PICTURE THIS MOVING IMAGE, Spike Island Studios, 40 Sydney Row, Bristol BS1 6UU. 0117 925 7010.

POZITIV PRODUCTIONS, PO Box 2646, Burghfield Mill, Reading, Berkshire, RG30 3YH. www.pozitiv.com 0118950 9050.

PRESENTER-LED PRODUCTIONS, PO BOX 17861, London W5 5ZP. www.presenterled.com 07074 737597.

RAVENSBOURNE COLLEGE OF DESIGN AND COMMUNICATION, Short Course Unit, Walden Road, Chislehurst, Kent BR7 5SN. www.rave.ac.uk/shortcourses 020 8325 8323.

REAL TIME, The Media Centre, 21 South Street Arts Centre, Reading, RG1 4QU. www.rael-time.org.uk 0118 901 5205.

SAFESETS TRAINING, Stable Studios, 15 Priest End, Thame, Oxon, OX9 2AE. www.productiondesign.co.uk 01844218072

THE SCREENWRITERS WORKSHOP, Suffolk House, 1-8 Whitfield Place, London W1T 5JU. www.lsw.org.uk 020 7387 5511.

SCRENE PRODUCTIONS, 126 Norwood Road, London SE24 9AF. www.screne.org 020 8671 7071.
SIF (SHEFFIELD INDEPENDENT FILM) 5 Brown Street, Sheffield, S1 2BS. 0114 272 0304.

UNIVERSITY OF SUSSEX, Centre for Continuing Education, Falmer, Brighton, BN1 9RG. www.sussex.ac.uk. 01273 606755

THAMES VALLEY UNIVERSITY, St Mary's Road, Ealing, London W5 5RF. www.tvu.ac.uk 020 8579 5000

TRANSITION MEDIA ARTS, Wellington House, 83 Main Road, Washingborough, Lincoln, LN4 1AY. Beehive.thisislincolnshire.co.uk/transitionmedia 01522 797207.

VFG LIGHTING, Pinewood Studios, Iver Heath, Buckinghamshire, SLO ONH. 020 8795 7000.

WARPAINT SCHOOL OF MAKEUP, 7 Hadleigh Close, Merton Park, London SW20 9AW. 020 8543 1996.

WEST HERTS MEDIA TRAINING CENTRE, South Way, Leavesden Herts WD2 7LZ. 01923 681602.

UNIVERSITY OF THE WEST OF ENGLAND BRISTOL, Media Works, Faculty of Art Media and Design, Kennel Lodge Road, Bower Ashton, Bristol, BS1 2JT. www.mediaworks.org.uk 0117 927 5881

WESTMINSTER KINGSWAY COLLEGE, Battersea park Road, London SW11 4JR. www.wesking.ac.uk. 020 7556 8000.

WFA MEDIA AND CULTURAL CENTRE, 9 Lucy Street, Manchester, M15 4BX. www.wfamedia.co.uk 0161 848 9785.

WOMENS INDEPENDENT CINEMA HOUSE, 40 Rodney Street, Liverpool, L1 9AA. 0151 707 0539.

YMTC (Yorkshire Media Training Consortium), 40 Hanover Square, leeds, LS3 1BQ. www.ymtc.co.uk 0113 294 4410.

YORK MULTIMEDIA NETWORK, Gateway House, 26 Swinegate, York, YO1 8AZ. www.ymn.net

Other courses

In addition to the above colleges offering courses in further education and short courses, there are a range of undergraduate and postgraduate courses on offer. These courses can be the BFI Media Courses UK, as listed, or through the BFI website.